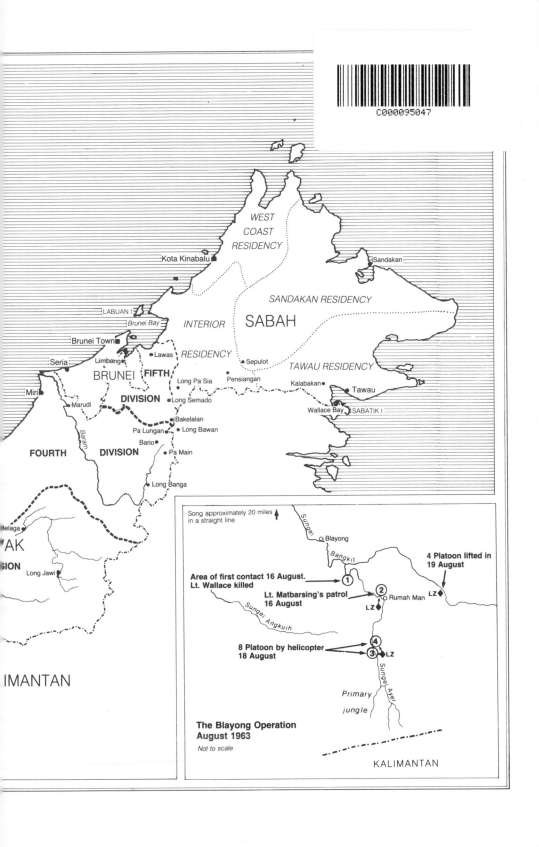

WEST
COAST
RESIDENCY

Kota Kinabalu

Sandakan

SANDAKAN RESIDENCY

INTERIOR

SABAH

LABUAN I

Brunei Bay

Brunei Town

Seria

Limbang

Lawas

BRUNEI

FIFTH

RESIDENCY

Sepulot

TAWAU RESIDENCY

Miri

Long Pa Sia

Pensiangan

Kalabakan

Tawau

DIVISION

Long Semado

Wallace Bay

SABATIK I

Marudi

Baram

Bakelalan

Pa Lungan

Long Bawan

Bario

FOURTH

DIVISION

Pa Main

Long Banga

elaga

AK

Long Jawi

ION

IMANTAN

Song approximately 20 miles
in a straight line

Sungei

Blayong

Bangkit

4 Platoon lifted in
19 August

Area of first contact 16 August.
Lt. Wallace killed

Lt. Matbarsing's patrol
16 August

Rumah Man

LZ

LZ

Sungei Angkuih

8 Platoon by helicopter
18 August

LZ

Sungei Ayer

Primary

jungle

**The Blayong Operation
August 1963**

Not to scale

KALIMANTAN

THE STEADFAST GURKHA

THE STEADFAST GURKHA

Historical Record of
6th Queen Elizabeth's Own Gurkha Rifles
Volume 3 1948–1982

by

CHARLES MESSENGER

A Leo Cooper Book
Secker & Warburg · London

First published 1985 by Leo Cooper in association with
Secker & Warburg Ltd, 54 Poland Street,
London W1V 3DF.

ISBN: 0-436-27780-8

Photoset in Great Britain by
Rowland Phototypesetting Ltd,
Bury St Edmunds, Suffolk
and printed by St Edmundsbury Press
Bury St Edmunds, Suffolk

DEDICATED TO ALL RANKS OF THE
6TH QUEEN ELIZABETH'S OWN GURKHA RIFLES
AND ITS FORBEARS WHO HAVE FALLEN
OR DIED IN THE SERVICE OF THE
REGIMENT

1817–1982

Contents

Foreword

by

FIELD MARSHAL THE LORD HARDING OF PETHERTON

It is a little less than forty years ago that I first got to know the 6th Queen Elizabeth's Own Gurkha Rifles when the Second Battalion of the Regiment, then forming part of the 43rd Gurkha Lorried Brigade, joined the 13th Corps, which at that time I had the honour to command, for the advance over the River Po to Trieste in the final stages of the Italian Campaign. Since then, I have got to know the Regiment well through the two years both Battalions served under my command during the anti-terrorist campaign in Malaya, my proud and happy ten years as Colonel of The Regiment, and numerous occasions since. Consequently I can speak with personal experience of some of the events and achievements recorded in this, the third volume of the History of the Regiment, which covers the period from 1948-1982.

It is a continuation of the epic story of over a century and a half of devoted and invaluable service in peace and in war to the British Crown, of courage and comradeship, of resolution and endurance under the most difficult conditions of climate and terrain in many different parts of the world.

There is a sentence contained, I believe, in the writings of the ancient Greek historian Polybius that reads:

> "Of all the forces that are of influence in war the spirit of the warrior is the most decisive one."

In my military experience extending now over some seventy years I know of no body of men who better embody the spirit of the warrior than the men of all ranks of The 6th Gurkha Rifles. It is not for nothing that they are known as "The Happy Warriors".

Like everyone who has ever served alongside this famous Regiment I would like to endorse the feeling by Brunny Short at the end of his vivid and entertaining account of a jungle patrol in Malaya when he wrote:

> "You can only thank your lucky stars for the troops around you, those magnificent qualities are just the same as they were when you first got to know them – maybe many years ago – and just the same as they will always be."

Gurkhas are true and worthy family men and no history of The Regiment would be complete without a fitting and specific reference to their womenfolk and children. My wife loved to visit them whenever she had the opportunity and always immensely enjoyed their warm smiling welcome and their happy company. It was always a great pleasure and encouragement for us both to see the splendid way in which they, like their menfolk, so rapidly adapted themselves to radical changes in their social, cultural and domestic environment without for one moment losing their natural modesty, dignity and charm. They made an outstanding contribution to the morale and contentment of the whole of the Regiment. It was most gratifying as well to see the way their children took keen and enthusiastic interest in and advantage of the available educational facilities.

I am immensely proud of my long and happy association with The 6th Gurkha Rifles, and count it the highest honour to have a further opportunity of offering all ranks my warmest congratulations on the brilliant achievements recorded in this volume, and my very best and most confident good wishes for continuing success in the future in all their undertakings.

HARDING OF PETHERTON
FIELD MARSHAL

April 1984

Introduction

I deem it a great honour to have been invited by the then Colonel of the Regiment Brigadier Sir Noel Short and Commanding Officer, Lt Col P. D. Pettigrew, to write this, the latest volume of the Regimental History of the 6th Queen Elizabeth's Own Gurkha Rifles. The two previous accounts, Volume 1 by Major D. G. J. Ryan and others (produced 1925) and Volume 2 by Lt Col H. R. K. Gibbs (1955), covered the periods 1817–1919 and 1919–1948 respectively. This volume takes the Regiment from transfer to the British Army in late 1947 up to the end of 1982 when it was about to leave Hong Kong for another tour in Brunei. The Malayan Emergency and Confrontation in Borneo are obviously the cornerstones of this period, as well as the uncertainty brought about by the withdrawal from Empire. Nevertheless, the Regiment, as the reader will see, has found itself engaged on many other varied tasks, and life is as busy now as it ever was. Indeed, 'flexibility' is probably the watchword of the Gurkha of the 1980s.

As a former Regular Officer in the Royal Tank Regiment whose service never took me East of Suez, I am very conscious of the fact that I have had no personal experience of much that I have written about. Nevertheless, the help given to me by the Regiment and my own experience as a military historian have given me confidence, and I earnestly hope that ex and serving members of the 6th Gurkhas will not be too disappointed with the result. In particular, I would like to thank Brigadier Brunny Short for his interest and encouragement, Lt Col Christopher Bullock, an old friend, who first mooted the idea when we were sharing endless watches on Exercise Wintex 1981, as well as giving me very useful information on his time in command, Major Duncan Briggs who acted as my "link man" with the Regiment from his desk in the Ministry of Defence, and Major

Gopalbahadur Gurung, who very generously gave of his time just as he was about to take the RMA Sandhurst Demonstration Company on a six-month tour to the Falklands. Brigadier Gil Hickey and Lt Col Roger Neath also very kindly read the manuscript and put me right on a number of points, and I am most grateful to them for their time and trouble.

As for the Regiment itself, Lt Col Paul Pettigrew very kindly invited me out to Hong Kong for Dashera 1982, and it proved a most memorable visit. The warmth and friendliness of all ranks, the opportunity to talk over old times with the *buroh sahibs* attending the Reunion, and the chance to observe the Regiment at work and play gave me a 'feel' which I would not have otherwise had. Apart from Paul Pettigrew himself, I would particularly like to thank Major (now Lt Col) Bob Richardson-Aitken, who arranged my programme, Captain Mark Harman, my "bear leader", Gurkha Major Jaibahadur Gurung for the seemingly endless hospitality of his Mess, and Brigadier Morgan Llewellyn who gave me a very detailed and informative briefing on the Gurkha Field Force.

Finally, I and the Regiment owe a very special debt to Colonel Hugh Pettigrew, father of the last Commanding Officer, for the vast amount of detailed research which he did for the History before I came to write it. Without his patient toil my task would have been many times more difficult.

London
1983
December

CHARLES MESSENGER

CHAPTER ONE

To a New Life

The beginning of 1948 found the Regiment at the climax of what had been a very unsettled period. It had begun in the middle of 1946 when orders had been given to move the Regimental Centre from Abbottabad, the home of so many generations of 6th Gurkhas, to Dehra Dun, where all the Gurkha Regimental Centres were being concentrated. There was, too, the turbulence brought about by the rundown of strength after the end of the war, which meant not just the release of many British Officers and Gurkhas, but also the loss of the 4/6th, which was disbanded at Abbottabad in early 1947. Although it had only been formed in March 1941, it had more than lived up to the reputation of the Regiment by its fine performance during the last year of the war in Burma.

There was, however, a more fundamental reason, which had produced a high degree of uncertainty throughout, not just the Gurkhas, but the Indian Army as a whole. This was the decision to grant independence to India. Everyone had been aware that this was to happen since early in 1946, but when Lord Mountbatten became Viceroy in March, 1947, he arrived with orders that independence was to be granted within six months, and the date was set for 15 August, 1947. This meant, of course, that the Indian Army was to be split between India and Pakistan.

Yet, behind the scenes, unknown outside the highest official circles until a week before Independence Day, lengthy and delicate negotiations had been taking place to transfer some of the Gurkha Regiments to the British Army.

In the Spring of 1946 Lord Wavell, the then Viceroy of India, had warned His Majesty's Government that the future Government of India would be unable to continue to employ all ten Gurkha

Regiments, and that their best forecast was four two-battalion regiments only. He recommended that HMG take steps to arrange for the residue of Gurkhas to be transferred on independence, but no action was taken until the following March when the six months deadline for independence had been set. During that first week of March, 1947, the Executive Committee of the Army Council considered the problem. On the one hand was the fact that the large wartime army was being run down, and one effect of this was the placing in "suspended animation" of many Regular British infantry battalions. Thus, to take on Gurkha battalions might well mean jeopardizing the future of many famous British regiments. But the arguments for the transfer were overwhelming. On the political side there was the need to maintain strong links with Nepal, and there was no doubt that the military connection was the best way of doing this. There was also the question of Malaya and, indeed, of British interests throughout the Far East. A division stationed in Malaya would provide an ideal reserve for the whole area, but the problem of finding British troops for it was very difficult in view of the Treasury-imposed manpower ceilings and commitments elsewhere. There was also the fact that it looked at the time as though the active term for the National Servicemen in the largely conscript post-war Army was to be one year only,* which meant that, when basic training had been taken into account, it was not worthwhile to post them to the Far East. A division of Gurkhas would solve the problem. Once the decision had been made, however, there was a need to act quickly before India ran down the Gurkhas.

First, the future Indian Government had to be approached to establish that they had no objection to HMG negotiating direct with the Government of Nepal. Then an agreement had to be made with the latter. Both these steps and the decision as to how many Gurkha regiments and battalions should transfer took time. The Tripartite Agreement for the employment of Gurkhas in the British and Indian Armies was not ratified until 9 November, 1947. In the end, it was decided that four two-battalion regiments should be nominated for transfer, and the announcement that the 2nd, 6th, 7th and 10th Gurkhas had been selected was finally made on 8 August. In a letter sent by the Adjutant General in India, Sir Reginald Savory, to the Commanding Officers of all Gurkha Battalions, some of the background was explained:

* It was Field Marshal Slim's first task on taking over as CIGS in November 1947 to get this term increased to 18 months.

"I know the decision as to which regiments of Gurkhas are to go to the War Office must have been a surprise to you and that you must be wondering how on earth the decision was made. I am, therefore, writing this letter that I may try to explain what may appear to you to be inexplicable.

The future of the Gurkhas has been the subject of prolonged negotiations at a very high level in which many were involved – the Foreign Office, the India Office, the War Office, the Government of India, the Government of Nepal, the High Commissioner in India for the United Kingdom, and last but not least, GHQ India. The negotiations have not only been very slow but also very delicate and were of a nature which made it quite impossible for me to give you a hint as to what was going on, because to have done so might have prejudiced the whole question whether or not any Gurkhas were to go to the War Office.

This delay has allowed us very little time between the date of the final decision and the 15th August. The War Office, when stating which particular regiments they would like, said that they wished the units to be spread as widely as possible over the recruiting area of Nepal. They specified the 7th or 10th Gurkhas and two regiments of Magars and Gurungs.* Among the last-named they particularly specified the 2nd Gurkhas owing to their close relationship with the 60th Rifles. The selection of the second regiment of Magars and Gurungs they left to us.

In making the final decision we had to be influenced by pressing problems of time and shipping.

Both the 7th and 10th Gurkhas have regular battalions in Burma. The 6th Gurkhas also have a regular battalion in Burma and we, therefore, chose them as our free choice for Magars and Gurungs. All these battalions were serving under the War Office (SEALF) [South-East Asia Land Forces] and their transfer from Burma to their ultimate home in Malaya would be a matter of arrangement by HQ SEALF.

The 2nd Gurkhas were the only Regiment which had no regular battalion overseas, but they were specially nominated by the War Office."

It had been a hard decision to make, and the disappointment of the regiments not selected was plain. Apart from having the 1/6th in Burma, the Regiment was probably also lucky in the influence that

* Those from western Nepal. The 7th and the 10th recruited from Eastern Nepal.

its Colonel, Field-Marshal the Lord Birdwood, could wield, as well as that of the Sixth's most distinguished serving soldier of the time, General Sir William Slim, who would shortly retire from being Commandant of the Imperial Defence College to work as Deputy Chairman of the Railway Executive of British Railways.

General Savory's letter then went on to explain what was to happen to the Gurkha officers and men of the four chosen regiments:

". . . every single one is being given the choice of whether he wishes to serve under the War Office, under the new Government of India, or either, or neither, so there is no question of forcing them to serve under one or the other. Whatever they do will be of their own free will. Those who serve under the new India will have the advantage of being eligible for regular commissions as fully-fledged officers, whereas those who serve under the War Office will not have that advantage."

However, they were expected to make up their minds by Independence Day, which gave them a bare week only. Not unnaturally the outcry was vociferous from the Regiments, particularly as terms of service had not been worked out, and the brief outline given that only five years' guaranteed service under the Crown could be offered, apart from the seemingly infinitely better prospects for advancement in the new Indian Army, meant that the majority of Gurkhas might understandably opt against British service. Nevertheless, the various battalions and centres did begin to carry out a referendum as ordered, and the results were encouraging. The War Office then relented, and the "opt", as it was called, was postponed until more precise details could be worked out.

One major problem that had not been appreciated at the time was that of the British Officers. They had been required to make up their minds as to their future by 1 July, 1947, before they knew that Gurkha Regiments were being transferred to the British Army. A number could stay on for a while as advisers to the new Indian and Pakistan Armies, but for the remainder there was the stark choice of transfer to the British Army, rebadged to British Corps and Regiments, or retirement. This added to the turbulence and also meant that many well-known faces began to leave the battalions and Regimental Centres. For the Gurkha, who has always been keen to serve under the British Officers whom he knows, this provided yet another reason for opting for India.

In order to resolve the problem of British officers to serve with the

Gurkhas, the Army Council decided in October, 1947, that they should be drawn from three sources. Ex-Indian Army regular and emergency-commissioned officers transferred to the British Army would be offered permanent service with the Gurkhas, while non-regular officers with previous Gurkha experience could volunteer for a limited number of Short Service Regular Commissions (SSRC). Finally, there would subsequently be the opportunity for other officers to be seconded to the Gurkhas for short periods.

A more ticklish problem was that of the Gurkha Officer (GO). He, of course, under the old Indian Army system, ranked as junior to all British officers. The original intention had been to offer GOs a British Army commission under the same terms. However, the new Indian Army was offering them exactly the same terms as their own officers, and hence it was understandable that they would not wish to join the British service. Something obviously had to be done to improve the terms, and that very same October, 1947 meeting of the Executive Committee of the Army Council, which had even gone so far as to seek legal advice on the matter, resolved that the King's Commission would be offered to all serving Gurkha Officers. But there was a rub. Since it was felt that it would still lead to difficulties if they were granted the full powers of a King's Commissioned Officer, it was decided that they should be given a Restricted Power King's Gurkha Commission (RPKGO), termed KGO for short. This, however, would not be broadcast, since this commission would be little more than they had before. Nevertheless, it was accepted that any Gurkha who achieved entry to RMA Sandhurst would be granted the full King's Commission, although it was thought that it would be some years before this could happen.

It was therefore not until December that the second opt could take place, and even then, as Major-General Whistler, who had been appointed to take charge of all Gurkha units transferring to the British service, wrote in a report dated 12 April, 1948:

"Result was disappointing and did not augur well for the recruiting drive which followed. However, recruiting exceeded all expectations and by the end of February it was confidentially [sic] expected that the target figure for 1 Apr 48 would be exceeded."

Indeed, when Headquarters British Gurkha Troops in India closed down at Delhi on 15 March, 1948, the totals of those who had elected to serve under HMG were as follows:

Opted	4288
Volunteers	452
Ex-soldiers	390
Local enlistment	92
Raw Recruits	2088
Recruit Companies	108
Yet to opt	732

This meant that all the battalions would move to Malaya with an unusually high proportion of recruits, who would need to be trained before the battalions could be considered fully operational.

Another problem was the decision that what was termed "The Gurkha Regiment" would be responsible for finding its own men for the supporting arms and services of the Gurkha Division. This was to be designated 17th Division, an apt selection in view of the distinguished service given by its predecessor in Burma, and in which many Gurkha units had served. This meant supplying the men from the existing Regiments, and, even more revolutionary, the 7th Gurkhas were told that they would be converted to artillery, a decision that was later rescinded.

During this time the 1/6th, under Lt Col Jim Robertson DSO OBE, were stationed at Rangoon, and declining numbers (they had fifteen British Officers and 330 Gurkhas only as at 1 January, 1948) meant that they found themselves very stretched in providing guards and on other duties. They left Burma on board HMT *Dunera* on 28 January, and arrived at their new home, Sungei Patani in Kedah State, Northern Malaya, three days later. However, they had a very much easier changeover than the 2/6th.

The 2/6th, commanded by Lt Col Freddie Shaw DSO, had been stationed in New Delhi since February, 1947. Initially its duties had been largely ceremonial, but it had become more and more involved in the communal unrest arising from Partition. When the initial opt was undertaken in August, 1947, they, together with the 1/6th, could boast of a 90% preference for service with the British Army. However, as the autumn went on and the uncertainty grew, many changed their minds, and when the second opt was taken in December, the Battalion was left with eight British Officers, three King's Gurkha Officers and 113 Gurkha Other Ranks only. On 1 January, the official day for the transfer to the British Army, the problems really began. For a start, there was the need to concentrate those opting for HMG into one company, but, at the same time, the non-opts were still under command. Also, the Battalion was still expected to provide the guards for Viceregal Lodge. Further headaches were the transformation of documentation from the old Indian Army to a temporary British Army system, the opening of a new Imprest Account and conversion to the British pay system, the handing over of all surplus Q stores to the new Indian Army, with the resulting closing down of all ledgers for audit, and finally the preparation of all accounts, both public and private, also for audit. To aggravate the situation still further, only two of the clerks had opted for the British service. In addition, the lack of NCOs was made more apparent by the order to send the nucleus of a training wing to Ranchi in order to receive new recruits. As the Battalion's Quarterly Historical Report said, it "proved to be quite a busy month".

February saw those opting for the Indian Army handed over, but this did not really ease the burden, because 192 ex-service recruits arrived, "who had to be clothed and equipped and for whom stores had to be 'begged, borrowed or stolen' as all ours had been handed over". Nevertheless, these men provided a much needed source of NCOs. It was also the month of packing up prior to the move to Malaya, and, since they were expected to be in Bombay by the end of the month, the lack of men for working parties was sorely felt. Finally, at the end of the month, having moved to Kalyan Transit Camp, Bombay, the Battalion received a draft of 180 brand new recruits. Then, on 3 March, the 2/6th sailed from Bombay to Singapore on RMS *Strathnaver*. Among others who came to see them off was the Nepalese Ambassador to India, who made a farewell address to the Gurkhas, which must have given them encouragement for their voyage to a new life.

This left 3/6th, of whom no mention has been made until now. Reformed in October, 1940, it took part in the first and second Chindit Expeditions, and during the latter Captain Michael Allmand and Rifleman Tulbahadur Pun had won the Regiment's two Victoria Crosses of the Second World War during the operations leading to the capture of Mogaung in June, 1944. Sadly, it was, of course, not possible for the 3/6th to join the British service, and, in December, 1947, it became the 5/5th Gorkha Rifles (Frontier Force) and passed into the new Indian Army. There was, though, consolation that this fine battalion would continue to live on, albeit under a new flag and a new name.

The main priorities on arrival in Malaya were to settle into the new environment and to train up the vast influx of recruits. The camp at Sungei Patani was in need of refurbishment and cleaning up, but potentially much better than the 1/6th had had in Burma.

However, the new home of the 2/6th, Wardieburn Camp, Kuala Lumpur, was frankly depressing, and the only saving grace was that the Battalion knew that it would not be there long as it was bound for Hong Kong within twelve months. Indeed, it was originally intended that the 2/6th would go direct to Hong Kong, but, because of its low strength, it was decided to send it to Malaya first. Yet hopes of a reasonable period of time in which to become activated and train the large numbers of raw recruits were shortly to be dashed.

16 June, 1948, marked the start of what was to be the twelve arduous years of the Malayan Emergency. On that morning three Chinese terrorists bicycled to Elphil Estate, twenty miles east of Sungei Siput, and murdered the manager, a Mr Walker. Almost simultaneously, two planters were also done to death in the same area. Indeed, one of the latter, an ex-8th Gurkha, had visited the Mess of the 2/2nd Gurkhas two days before and voiced his concern that trouble was brewing. On 17 June the Government declared a State of Emergency in Perak and Johore, and next day this was extended over the whole of the Federated Malay States.* However, the events of the 16th were not an isolated outbreak of violence, for this had been simmering for some time, and the British troops in the area had been engaged on anti-guerrilla operations since the previous year. The causes of the violence go back even further.

The root of the problem lay in the spread of Communism in South-East Asia, which began some thirty years before with the Second Congress of the Communist International in the summer of 1920. Two years later the First Congress of the Toilers of the Far East was held in Petrograd and was attended by representatives from the area. At that time Malaya was not regarded in Moscow as a "colonial possession", and this was correct as Malaya had never been a British colony. The nine Malay states had, in return for agreeing to British protection and the installation of British Residents in each to advise the Sultan, traditionally enjoyed a fair measure of autonomy. Nevertheless, even though the USSR might not be interested, the Chinese Communist Party was, particularly after the death of Sun Yat-sen in 1925. The focus of attention was Singapore, especially since the South Seas Branch of the China Communist Party was formed there in order to represent the in-

* At this time Malaya was divided administratively into three groups – The Federated Malay States of Selangor, Perak, Pahang and Negri Sembilan, the Unfederated Malay States of Johore, Perlis, Kedah, Kelantan and Trengganu and the Straits Settlements of Malacca, Penang and Singapore.

terests of both Malayan and Dutch East Indies party members. However, police raids severely restricted its activities and it lasted only five years. A General Labour Union had been formed under the same umbrella shortly afterwards, and did prove more successful, but in April, 1930, it was decided to start afresh, and a Malayan Communist Party and Malayan General Labour Union were created. They were answerable to the Pan Pacific Trade Union Secretariat in Shanghai, which itself came under Comintern's Far Eastern Bureau. A member of the French Communist Party was sent to Singapore to help with the reorganization, but he was arrested shortly afterwards, and this, together with Ho Chi Minh's arrest in Hong Kong, brought about the demise of the Far Eastern Bureau.

The Malayan Communist Party (MCP), without the necessary direction from above, now went through a period of isolation, but was able to survive and indeed grow as a result of the world economic depression and the Japanese invasion of China. The former enabled the Party to exploit grievances among the Chinese work force in the country, while the latter, through the formation of what were called Anti-Japanese National Salvation Associations, provided an emotive rallying point. Throughout, the Party's main aim remained that of ridding Malaya and Singapore of British imperialism and instituting dictatorship by the proletariat.

When war broke out in Europe in 1939 the MCP took the international Communist line that it was a conflict between capitalist states and of no concern to the Communists, a policy which would switch dramatically with the German invasion of Russia. However, the Japanese invasion of Malaya in December, 1941, suddenly brought the MCP, which had been an illegal organization up to then, official acceptance. Its offer of assistance to the Government was gratefully accepted, and it provided a number of recruits for the 101 Special Training School, which had been hastily set up under Lt-Col Spencer Chapman to provide cadres for special operations against the Japanese. In the event, the Japanese advance proved too quick for much to be organized, and Spencer Chapman's men were forced to take to the jungle, where they remained for the rest of the war. However, those groups which contained MCP members did link up with the local party cells and in this way an infrastructure was created in order to carry on the fight against the Japanese. During the remainder of the War the MCP was severely harassed by the invader, and many of its members were killed or arrested, but it says much for its spirit and determination that it was able to survive. It was helped in this by three major factors. Firstly, the Malayan

People's Anti-Japanese Army (MPAJA), the military wing of the MCP, began, in the later years of the war, to receive substantial help from Force 136, which was the element of South East Asia Command (SEAC) responsible for the co-ordination of operations within Japanese-occupied territories. Large quantities of arms and ammunition, along with liaison officers, were parachuted into the jungle, and this resulted in a substantial arms build-up. Secondly, although only a small proportion within the groups themselves were MCP members, the long periods of inactivity in their jungle camps enabled them to work on their brother guerrillas and this led to a swelling of the membership. Finally, the fact that the Japanese had overrun Malaya and taken Singapore so easily seemed to prove that the British were fallible, and there was a belief that they would be hard put to it to re-establish themselves in the country after the war. In February, 1943, the MCP issued a nine-point anti-Japanese Programme, which confirmed its primary aim of driving the Japanese, whom it termed as "fascists", out of Malaya and establishing a Malayan republic, and, in pursuance of this, it agreed to subordinate itself to SEAC for military operations.

With the Japanese forces defeated, it was necessary to retrieve the arms and equipment passed to the indigenous bands operating in the jungle and, as an enticement, each guerrilla was offered the sum of 300 Malayan dollars and a campaign medal if he would surrender his weapons. However, the Communist groups paid little heed to this and consequently had a large arsenal built up and ready for the moment to strike.

Immediately after the war the British attempted to overhaul the system of government, which was in a shaky state. Casualties suffered during the war by those who had been in the administration, resentment by those who had stayed and endured the Japanese occupation towards those who had spent the war elsewhere, and the fact that many Chinese had migrated to the fringes of the jungle, forming communities which were difficult to govern and were hotbeds of Communist influence, had all contributed to this. To resolve this, the Attlee Government attempted to form a Malayan Union which would give the Chinese and Indians the vote to which they had not been formerly entitled, and reduce the nine Sultans to mere figureheads. However, the Malays, perceiving that the Chinese might become too influential, objected. So the concept was short-lived and in late 1946 the Colonial Office thought again. The result was the Federation of Malaya which came into being on 1 February, 1948, enabling the Sultans to retain some of their power, and the

Malays to continue to have a major say in the running of the country.
The Chinese remained virtually as isolated as before.

In the meantime Chin Peng, the leader of the MCP, had formed
the Malayan Peoples Anti-British Army, changed in 1949 to the
Malayan Races Liberation Army (MRLA), which was built largely
around the wartime guerrillas. Under the Central Executive Committee were three regional bureaux, covering north, centre and
southern Malaya. These supervised state committees, who controlled district committees, numbering fifty in all, who ran the lowest
element of all, the branch committees, of which there were usually
four per district. The committees controlled both political and
military affairs, and under each state committee was a regiment,
which consisted of a number of companies and independent platoons. As a result, at the outbreak of the Emergency there were
upwards of 10,000 armed and well-trained guerrillas. In addition,
there was a logistic element in the Min Chong Yuen Tong or People's
Movement, more commonly known as the Min Yuen. They were not
members of the Communist Party, but sympathizers, with responsibility for keeping the guerrillas supplied with food, medicine and
other needs, as well as being called out for particular operations.
They lived at home, liaising with the local branch committee.

In February, 1948, as the new constitution for Malaya came into
force, an MCP delegation attended the Communist Asian Youth
Congress in Calcutta, and, inspired by the ferocity of some of the
speeches there, Chin Peng drew up his final plan of campaign. This
would be in three phases. First, the MCP would establish liberated
areas around their jungle camps. Part of this phase involved the
elimination of the few Europeans living in these areas. These areas
would join together to become a "Liberated Country", and finally
the populated parts would be attacked, leading to a breakdown of
law and order, disruption of the economy and the establishment of
a Communist government. In the event, the plan was triggered off
by a Radio Malaya broadcast on 6 June, 1948, when Malcolm
MacDonald, Commissioner-General for South-East Asia, warned of
the Communist campaign of intimidation in the plantations, mines
and factories. This had been growing over the past two months and
had been conducted by gangs known as the "Blood and Steel
Corps". Even though troops had been assisting the Police in curtail-
ing their activities, it was not until Malcolm MacDonald's speech
that it was generally recognized that there was more than criminal
intent involved. Chin Peng decided that he could wait no longer and
he and his men took to the jungle.

CHAPTER TWO

Emergency in Malaya – The First Phase

As soon as the State of Emergency was declared, both Battalions found themselves committed on operations. For the 1/6th, this meant deploying companies to Alor Star, Kulim and Penang, and extensive anti-terrorist patrolling in Kedah. Battalion HQ remained at Sungei Patani. One problem was that the low strength of trained soldiers required recruits to be employed on static guard duties, and their training was thus brought to a halt. For the 2nd Battalion the situation was much the same, with static guards and joint patrols with the police being mounted in Selangor. Indeed, it was during this period that the first Gurkha casualty of the Emergency occurred, when a patrol under Captain (KGO) Hiralal Gurung* was ambushed at Kajang and Lance Corporal Dhawa Ghale was killed. A few days after this, orders were received to move to Pahang and help re-establish the police in a village called Gua Musang, whose police station had been attacked and overrun and a Malay Regiment platoon ambushed in an attempt to retake it, suffering four killed, including a British Officer. With a 1/7th Gurkha company and another from the Malay Regiment under command, "Shaw Force", as it was designated, after Freddie Shaw, set off on foot, for there was no road and the railway was not running, to cover the seventy miles to the village. They succeeded in restoring law and order in Gua Musang, and then moved westwards into the Pulai area to drive out the terrorists whom it was believed had taken refuge there. Clearing the area, with its numerous small rivers, caves and thick jungle occupied the 2/6th for the next month. Supported by RAF Spitfires with cannon and rockets and resupplied almost entirely from the air, the Battalion succeeded in its task, but not without cost. Although

* Father of a future Gurkha Major, Maj (QGO) Jaibahadur Gurung MVO.

one terrorist was known to have been killed and a sizeable quantity of arms and ammunition, including an anti-tank gun, were captured, losses were one rifleman killed and one wounded in an ambush.

Nevertheless, as General Sir Neil Ritchie, C-in-C Far East Land Forces, quickly realized, the best way of breaking the back of the terrorist organization was constantly to harry it with offensive operations like this, rather than remain tied down on the defensive. Another early means of pursuing this concept was Ferret Force. This was raised in early July as a special force designed to operate in small groups with the object of tracking the terrorists down in their camps and destroying them. It owed much of its concept to the Force 136 experience of the Second World War and indeed was commanded by an ex-member, John Davis. Another experienced jungle fighter, Lt Col Walter Walker, who would later command the 1/6th, was sent to train and equip it. Battalions were ordered to contribute to it, and the 2/6th sent twenty experienced men under Captain "Trigger" Tregenza. Although they were disbanded at the end of the year, the experience gained by these long range patrols was of much value to the conduct of operations as a whole, but the requirement to lend experienced riflemen at a time when they were at a premium was sorely felt by the Battalions.

The 1st Battalion also suffered its first fatality early on, when Rfn Gopilal Gurung was killed and two others wounded in a brush with terrorists on 26 July during a sweep operation in conjunction with the 2/2nd Gurkhas, police and armoured cars of the 4th Hussars in the Grik-Lenggong Valley. For the latter it was their first taste of jungle operations, having only just arrived in the country, with 2 Guards Brigade, as the first reinforcements sent out to deal with the Emergency. Further operations, including the evacuation of Chinese squatters – always a likely source of food for the Communist terrorists – followed, culminating in a screening operation of some 700 squatters, of whom eight were positively identified as bandits and arrested.

Dashera that year, the first in the British Army, was celebrated in early October, and took the form of the many that followed during the twelve years of the Emergency. Neither Battalion was able to concentrate everyone back in barracks, but nevertheless every member of the Regiment had at least one party, whether it was Kalratri or a company *bhoj*.* Both Battalions then went back on operations, but

---

* For those unfamiliar with Gurkhali and Malay words used in this book an explanatory glossary appears on p. 119.

the 2/6th shortly received the news that its move to Hong Kong was now confirmed and that it would take place before the end of the year. It was placed under command of the Guards Brigade, who would eventually take over the area, and was stood down just before Christmas in order to prepare for the move.

A significant event occurred in December when Major (KGO) Kajiman Gurung attended a board to pass the necessary tests for his commission as a Gurkha Commissioned Officer (GCO). He was successful, and now had the same authority as British officers with the Gurkhas, the equivalent of the former Gurkha Officer now being a King's Gurkha Officer (KGO). Pahalman Gurung of the 1st Battalion was made a GCO at much the same time.

The 2/6th left Kuala Lumpur on Boxing Day, having stood down from operations on 15 December. They went by train to Singapore and embarked in HMT *Dilwara*, arriving at Hong Kong on the evening of New Year's Day, 1949, disembarking on 2 January. They then went by special train to their camp at San Wai in the New Territories. The camp itself was still in a rather primitive state, being little more than a collection of Nissen huts, which were in a poor state of repair. Nevertheless, the intention was to build a permanent camp here during the next two years and, in any event, compared with the tents of Kuala Lumpur, it was almost luxury, especially for the Gurkha families. The first task was administration. With companies deployed away from Battalion Headquarters for most of the time in Malaya, documentation had suffered, especially with the continuing shortage of clerks. This took three weeks to sort out, and then there was pressure to get on with individual training, which again had suffered from operational commitments in Malaya, before collective training in the Colony began.

At this time the civil war in China had been virtually won by the Communists under Mao Tse-tung, but there was little perceived threat to Hong Kong. Nevertheless, with the spectre of the Japanese invasion of December, 1941, still present, defence of the Colony against external aggression was the prime role of the Garrison. A more immediate problem was that of armed bandits, who occasionally came across the border into the New Territories, and one platoon was on permanent stand-by to assist the police in dealing with them, although, apart from one false alarm, it was never called out. There was also a significant guard commitment for the Governor, the GOC and the Supply Depot at Shamshuipo, and even, for a time, on two psychiatric cases in BMH Hong Kong. All these interfered with individual training, but the staff proved most understanding and helpful.

The situation changed dramatically in April, 1949, when HMS *Amethyst*, protecting British interests on the Yangste River, was fired on by Communist guns, suffered casualties, and, for a time, until her epic dash for the sea, was trapped. The possible Chinese Communist threat to Hong Kong became very real, not only from without, but also with the danger of unrest within the local Chinese population. In addition to the guards mentioned above, the Battalion now also had to provide troops to reinforce police stations on the frontier, carry out "showing the flag" patrols on the Border by day and have a platoon on thirty minutes stand-by for internal security duties. There were also numerous brigade schemes designed to practise rapid deployment to defensive positions back from the Border. Moreover, with the introduction of anti-tank guns and carriers into the battalions, the deployment positions had to be resited. Routine operational duties tied up two rifle companies, but the manpower problem was aggravated by the leave parties sent to Nepal. The first had left Malaya in November, 1948, and was not due back until July, 1949, while the second was scheduled to leave in April, thus leaving a crucial gap of three months at a critical time. It was therefore decided that the Battalion's share of recruits from the Regimental Training Wing in Malaya should be sent to Hong Kong as an immediate reinforcement in case of emergency. It was, however, to be kept as a separate company in order to carry on with its training, although on arrival two days before No 2 Leave Party was due to depart, Colonel Shaw immediately absorbed fourteen recruits, who were re-enlistments, into the main body of the Battalion. Thus the shortage of manpower was still sorely felt. Indeed, as at 30 June, 1949, the 2/6th could only muster 510 all ranks against a posted strength of 808. The result was very little time for leisure pursuits, and sport suffered severely, with only the soccer team carrying out any form of training.

The 1st Battalion, meanwhile, was kept at full stretch on operations. There was little to show for their hard work, but at least continual patrolling and sweep operations kept the terrorists at bay. Then, in January, 1949, tragedy struck. A and D Companies were operating on the Malaya–Siam border, and Ronnie Barnes had just taken over the former from Bill James, so that he could prepare for and take the Staff College exam. On the 13th Barnes was patrolling with one of his platoons when his lead scout saw terrorists coming down the path towards him. He killed one with his Sten gun, but the platoon was now engaged by a burst of fire. Ronnie Barnes and the leading section were hit, Ronnie being killed instantly. Tulparsad Pun, the platoon commander, was mortally wounded shortly afterwards. The middle and rear sections returned fire, and also suffered casualties, including Lance Corporal Gaine Gurung, who, although wounded in the right shoulder, seized a Bren gun from a wounded rifleman and kept up accurate fire for almost half an hour until he too was killed. For this he was later awarded a posthumous Mention in Despatches. Eventually, the two sections managed to extricate themselves with their wounded. One rifleman, however, who had been hit three times, could not get away and was still there when the terrorists moved in to collect arms, etc. One fired at him, but it was a misfire. He feigned death, waited until nightfall and walked the four miles back to the base camp. In all, the Battalion lost eleven men killed and six wounded in this action, which was a very cleverly laid ambush, and indicative of the good intelligence which the terrorists had. The loss was deeply felt and signals of sympathy from General Ritchie and the Sultan of Kedah were much appreciated. Indeed, the latter attended Ronnie Barnes' funeral at Alor Star next day. As for Barnes himself, he was typical of the British officer serving with the Gurkhas at the time. After wartime service in the 2/6th and 4/6th he had, like many others in the uncertainty of 1947, reverted to the British Army and gained a permanent commission in the Essex Regiment. He then volunteered to return to his old regiment, and had only been back with the Battalion for a fortnight. He would, however, be remembered for a long time to come for his great natural charm, cheerfulness and enjoyment of life.

At this time the accent was still on large operations, and there were two main types. In the first, the idea was to surround an area in which terrorists were suspected to be and then send in a force to flush them out and kill them, while the other method was to drive the guerrillas on to a prearranged line of ambushes. Typical of the second type was Operation NAWAB (8–18 March, 1949), which involved the 1/6th, 1 KOYLI, three troops of the 4th Hussars and the Police. While the KOYLI "beat" through the area, and the 1/6th acted as stops behind them, the 4th Hussars and Police set up a series of ambushes. Only one bandit was killed (by 1 KOYLI) and three captured, but fifty suspects were held, and the Battalion destroyed six camps, seizing documents and clothing. Nevertheless, it was a frustrating business and there was a growing feeling that the more elaborate the operation, the less likely any success. Indeed, voices began to be raised against these large operations in favour of longer-range smaller patrols acting on firm intelligence. However, there was some doubt at this time about how long soldiers could effectively operate in the jungle without a break. Most operations lasted little longer than ten days, and there was little recent experience, apart from the more conventional jungle operations of the Burma Campaign, on which to base any workable policy. Indeed, a 1/2nd Gurkha view at the time was that they could operate "flat out" for ten days, but then needed a two-day rest. When, during Operation PINTAIL I (July–August, 1949) one 1/6th company remained in the jungle for twenty-seven days and easily broke the Battalion record up to that time, it was noted that, although the men were none the worse physically, they had lost some of their alertness. In fact, the general pattern of rest at this time was for one company at a time to be back in Sungei Patani, usually for three to four weeks, where it also carried out routine guards and duties, while the other three rifle companies, and often the Battalion Tactical Headquarters, were out on operations.

Even so, although few tangible results had been achieved, the Battalion was cheered by the publication of the first Honours and Awards list of the Campaign on 8 April, 1949. It included an MBE for Captain (KGO) Manu Gurung MC and seven Mentions in Despatches, including Gil Hickey, Derek Organ and Tony Taunton. The 2/6th, too, were delighted with the OBE conferred on Freddie Shaw for the operations of Shaw Force, as well as an MBE for Captain (KGO) Nandalal Thapa and six Mentions. Further cause for celebration in the 1st Battalion was the winning of the first playing of the Nepal Cup for football, now the premier sporting trophy of the Brigade of Gurkhas. Trained and managed by Lt Jimmy Lys* and Major (KGO) Lachhiman Gurung, the Gurkha Major, who both played in the team, victory in the final was eventually gained over the 2/7th after going into extra time, with the score at 3-2. Thanks to North Malaya District and 1st Bn KOYLI, who very kindly organized relief detachments and arranged for the Battalion to be stood down from operations, everyone was able to gather back in Sungei Patani for Dashera, which was celebrated in traditional style. It was also at this time that General Sir Neil Ritchie handed over as C-in-C Far East Land Forces to General

---

* His son George followed in his footsteps by playing in the 1983 Nepal Cup team.

Sir John Harding, who was later to become Colonel of the Regiment.

Immediately after Dashera, the Battalion was deployed to Pahang, where it spent four months on prolonged anti-terrorist operations, under command first of 2 Guards Brigade and then of 48 Gurkha Infantry Brigade. The Commander 48 Gurkha Infantry Brigade was Brigadier Osborne Hedley, who had won three DSOs in Burma, eventually commanding 26th Indian Division just before the end of the war. A 5th Gurkha himself, he took much interest in the 6th since his own regiment had also had its Regimental Centre at Abbottabad. Later he became GOC 17th Division and Major-General Brigade of Gurkhas. The Gurkha himself will perhaps remember General Hedley best for his introduction of "duty free" rum. Unfortunately the move to Pahang meant that most in the Battalion missed the visit to Sungei Patani by the CIGS, Field Marshal Sir William Slim, the Regiment's most distinguished ex-member, who had fought alongside the Battalion in its epic action at Sari Bair in 1915, and then had joined the Regiment on transfer from the British Army in 1920, serving with it until he went to command the 2/7th in 1938. He spent the night at Sungei Patani, but, alas, there was no one present who had actually served with him in the Regiment. Nevertheless, he was obviously delighted to see once more "Bubble and Squeak", the two 7 pounder mountain guns which had stood outside the Mess at Abbottabad, and told amusing tales of how he and others used to carry out gun drills on them during guest nights.

The operations in Pahang were arduous, but although the Battalion suffered casualties, there were some successes. 'Fairy' Gopsill's C Company succeeded in killing six terrorists in an attack on their camp in the Jerantut area of Eastern Pahang, which gained him the DSO to add to his MC won in French Indo-China and previous Mention in Despatches in Malaya. Unfortunately, shortly afterwards he was lost to the 7th Gurkhas on transfer. Another operation during this period also involved C Company, commanded this time by Bill James and operating as part of a six-company force with Philip Townsend in overall command. Again it was in the Jerantut area. A Battalion account describes it vividly:

"Whilst rations were being issued, the Company Commander reconnoitred the tracks left by the enemy. They were fresh and obviously made that day. The strength of the bandits had previously been estimated at from 150 to 200, and this was at once confirmed by the track they left through the rubber estate. The company at once set off in pursuit. Presently the trail was lost, at a spot where it issued from the estate and entered a squatter area, but the men cast around and in about half an hour discovered the path again, running into hill jungle. The pursuit was taken up again until dark, when camp was made at about 6 p.m. It was estimated that the enemy were some three or four hours ahead, and that, by making an early start, we would come up with them in the course of the next day. During the night there was nothing to report from our patrols. To understand what took place the next day it is necessary to know something of the camp the pursuers were to come across. It consisted of about thirty huts on a reverse slope running down to a large swamp. The huts were set in thick jungle, traversed by tracks leading to actual huts. On the right flank, our right, there was a steep ridge covered with dense thorn jungle. The left flank was thick creeper-matted jungle. These details were not, of course, known to us beforehand.

The pursuit was taken up again at 6.15 a.m., and continued until late afternoon with nothing special to report. The first intimation of the enemy's presence was a burst of Sten gun fire at our leading scout, fired by a bandit sentry sitting astride a log across the path. The scout returned the fire, dropping the sentry, and 8 Platoon charged to their front down the slope whilst the Company Commander shouted to 7 Platoon to take the right flank and 9 Platoon, the left. Firing broke out on all sides. 8 Platoon with Company Headquarters found themselves engaged by twenty to thirty bandits in the actual camp. Six bandits fell here and the rest did not stand up to the charge, making off under the cover of Sten, tommy-gun and rifle fire directed at 8 Platoon from huts further down the slope. Meanwhile, three Brens and other weapons opened up on 7 Platoon on the right flank from the spur which they were attacking; and short bursts of fire could be heard on the left where 9 Platoon were advancing. Fortunately, most of the fire was high, for 7 Platoon, in particular, were having a difficult time making their way up the steep slope, with its dense thorn jungle, in the dusk. . . . About twenty minutes had passed from the first shot when a bugle call sounded. This was the signal for a terrific outburst of fire on 7 Platoon from three Brens with Stens, tommy-guns and rifles, under cover of which a party of about thirty bandits charged wildly down on them screaming and shooting as they came. They got within about thirty yards; numbers were hit and fell, but were dragged away in the thick jungle under heavy covering fire. Another bugle sounded, and a similar charge was delivered against 8 Platoon in the camp, and was similarly repulsed, the screams of the wounded indicating a number of casualties. The advance was then ordered, but 8 Platoon became bogged down in the swamp and were therefore directed to hold to a line whilst action continued on the flanks. . . . Firing now became sporadic and died away completely at the notes of a third bugle call. At the same moment, heavy firing broke out where 9 Platoon had gone, on the left, and sounds of a Gurkha charge were heard. What had happened was this. The commander, Captain Dalbahadur Limbu, had lost contact with the rest of his platoon in the half-light and found himself with three men confronting about thirty bandits on a small path at the edge of the swamp. At that moment the bandit bugle blew. Calling to an imaginary company to charge, Dalbahadur and his three men raised the battle cry and charged the enemy, putting them to flight with a loss of 2 killed and 6 wounded.

This was the end. The pursuit was taken up and meanwhile the camp was searched and yielded arms, ammunition, food and the loot of previous outrages. Blood was everywhere, but only four bodies of bandits were recovered, although a careful check showed that twenty to twenty-five had fallen. We had no casualties and had fired only 700 rounds of all natures, compared to heavy expenditure by the bandits. The further pursuit is another story."

By the time the Pahang operations were finished for the 1st Battalion, the 2/6th was about to rejoin them in Malaya. The previous year had continued busy for the 2nd Battalion, beginning with a move in June, 1949, from San Wai Camp while it was rebuilt to a tented camp two miles away (the families had all been moved to Kowloon a month before). The threat from across the border had led to the formation of 40th Division, of which the Battalion became part as member of 26 Gurkha Infantry Brigade. Throughout the rest of 1949 the 2/6th was kept busy with operational commitments and exercises. One particular problem was that the defensive layout on the border was constantly being changed, which meant much digging. Still, the manpower situation improved and by August, 1949, the Battalion could boast of a posted strength of over one thousand men, including the reconstituted Pipes and Drums, now resplendent in white cutaway tunics and shorts, with rifle green plaid, kilmarnocks, tartan hose in black, green and red with white spats, and black waist and crossbelts. In March, 1950, with great sadness, the

Malaya – 1/6th Gurkhas patrol base

Malaya – 2/6th GPMG gunner

Malaya – 2/6th wireless operator

Battalion bade farewell to Freddie Shaw, who left after five years in command of battalions of the Regiment, first the 3/6th and then the 2/6th. He was succeeded by Ralph Griffith who would take the Battalion back to Malaya.

The 2nd Battalion arrived in Singapore in April, 1950, and was promptly deployed to Kluang in central Johore, and immediately began operations, the companies being based in various rubber estates. Another to arrive at the same time was General Sir Harold Briggs, who had been called out of retirement to become Director of Operations Malaya, with the task of planning, directing and co-ordinating the police and military operations against the terrorists. Within a few weeks he had formulated what became known as the Briggs Plan, which strongly influenced the conduct of future operations. In essence its concept was to clear the country step by step from south to north. Firstly the urban areas would be secured and confidence among the local population built up so that information would be freely given. Then the Communist organization within these areas would be broken up, and the terrorists deprived of their support. Finally the terrorists would be destroyed by forcing them to attack the Security Forces on their own ground. Briggs considered that the existing civil administration was unsuitable for dealing with the problem in that it did not provide for the close links needed between police, army and the civil government. Thus, at Federal level the Federal War Council was set up; each state had a War Executive Committee and at District level there was the District War Executive Committee, which was built around the local battalion commander, senior police officer and district officer. The Federation Army and police were enlarged, and a Special Constabulary and Village Home Guard were brought into being. Initially, with the arrival of two brigades' worth of reinforcements, including the 2/6th, Briggs hoped that his first operations in the south would bring about immediate results, but this was not to be, and it soon became apparent that it could only work in the long term. As he himself wrote:

"The problem of clearing Communist banditry from Malaya was similar to that of eradicating malaria from a country. Flit guns and mosquito nets, in the form of military and police, though giving some very local security if continuously maintained, effected no permanent cure. Such a permanent cure entailed the closing of all the breeding areas. In this case the breeding areas of Communists were the isolated squatter areas, the unsupervised labour on

estates, especially smallholders and Chinese estates without man-
agers. Once these were concentrated there might be some chance
of controlling the Communist cells therein. This showed clearly
that a quick answer could not be reckoned on."*

Meanwhile the 1/6th had spent the early part of the summer back
at Sungei Patani on retraining. The pace of operations and the fact
that battalions were split up, often over a wide area, meant that
training certainly, and discipline possibly, could suffer. Hence, from
early on in the Emergency, each battalion was allowed a period of six
weeks in order to carry out annual classification on its weapons, put
polish on anti-terrorist drills (from 1952 onwards these were based
on an official manual *The Conduct of Anti-Terrorist Operations in Malaya*,
known as CATOM or the Atom Pamphlet) and drill, update
documentation and carry out general administration, as well as
giving the opportunity for sport and local leave. This is not to say
that companies did not carry out their own training when they had
the opportunity, and the main emphasis of this was always on
shooting. General Templer, writing in the Foreword of the first
edition of CATOM, stressed ". . . the vital importance of accurate
and quick shooting, particularly with single shot weapons. If only we
can double the rate of kills per contact, we will soon put an end to the
shooting in Malaya."

At the end of retraining in July, 1950, the 1/6th were sent south, as
part of the Briggs concept, to the Bahau area of Negri Sembilan,
leaving a rear party at Sungei Patani. Here they were, like the 2/6th
in Johore, deployed in company bases on the rubber estates. There
was little intelligence to be had on arrival since their predecessors, a
battery of 26 Field Regiment RA, had been mainly involved in
railway protection. Constant patrolling was the order of the day, and
contacts were few to begin with. During the autumn, however, there
were a number of minor successes, but on the whole it was
an arduous and frustrating time. In terms of personalities, it was
with sadness that the Battalion said farewell to one of its major
"characters". Captain Pahalman Gurung SB OBI MBE had joined
in 1919 and was somewhat unusual in that he was an eastern
Gurkha, whereas the Regiment's recruiting area had always been
firmly in the west. Before the war he was one of the few Gurkha Drill
Instructors at the Indian Military Academy at Dehra Dun, and "his

* *Report on the Emergency in Malaya* April, 1950 to November, 1951, (Kuala
Lumpur, 1951).

twelve tunics and fourteen pairs of board-like shorts were a subject of wonder even in that age of elegance". During the war he was Subedar Major, and then one of the first GCOs of 1948.

The 2/6th remained at Kluang during this time carrying out much the same programme and experiencing the same frustrations as the 1st Battalion. One piece of history was made when a Pun from B Company was the first casualty to be evacuated by helicopter from the jungle. This was in September, 1950, and much publicity was given to the occasion, although the reason was merely one of toothache. Dashera was also interesting in that the RAF took over responsibility for the area using air control methods with their Lincolns and Brigands while the companies celebrated in their respective bases. Then, in December, came an unexpected break in the routine operational pattern. The cause was the Maria Hertogh riots in Singapore. She was a child of Dutch parents, who had fled Indonesia at the beginning of the war, leaving her in charge of a Malay woman who had brought her up in the Muslim faith. When, at the age of thirteen, a marriage was arranged for her with a much older Malay, her mother heard of it and smuggled her out of the country. She was brought to Singapore where her case was heard before the Chief Justices' Court. By this time Muslim anger was intense, and there was a very real danger that the fires would spread to Malaya and undo much of the good work in improving relations among the various communities which the Briggs Plan was slowly bringing about. The Battalion, under the temporary command of David Powell-Jones, was alerted and ordered to Singapore immediately. The problems of getting outlying sections and platoons in were immense, and at least one company commander had to resort to mortar fire in order to persuade his platoons to open up on the wireless link. Nevertheless, the move went without a hitch and the Battalion found itself involved in anti-riot drills in an urban area. Luckily, it did not have to shoot in anger, although in other parts of the city fire had to be opened. Apart from supporting the police, in Powell-Jones's words:

"There were patrols to extricate Europeans and Eurasians from isolated areas, patrols to cover the clearing of wrecked and burnt-out vehicles, patrols to enforce the curfew. There were flag marches. And there was waiting, a lot of waiting; waiting in reserve; waiting outside the Mosque to ensure that peaceful atmosphere for Friday prayers that the Koran enjoins; waiting for the dawn after the rude awakening from an uneasy sleep by the cry

of the muezzin, electrically amplified from each (praise Allah!) of
the four minarets high above – and there were the administrative
headaches: billetting; how and where to cook in a great city; how
to get food to scattered detachments; how to keep the men looking
clean."

The unrest soon died down and the Battalion returned to Kluang
surprisingly refreshed by the "holiday".

As far as the Brigade of Gurkhas as a whole was concerned 1950
brought further changes, but, rather than being unsettling, they
were to the good in helping it to establish itself within the British
Army. For a start, it was originally intended that the Permanent
Cadre of officers would waste out and that the Brigade would then
rely on seconded officers and Sandhurst-trained Gurkha officers.
This was now altered, and a Permanent Cadre re-established for the
infantry element. Officers could join it after one year's secondment
or by direct commissioning from Sandhurst, and the first of the latter
category to be commissioned into the 6th Gurkhas were Colin Scott
and Desmond ("Demi") Walsh, whose names will appear elsewhere
in these pages. All young officers were to do an attachment with their
affiliated British regiments in Europe, the Rifle Brigade in the case of
the Sixth, prior to joining the Gurkhas. It was also agreed that
selected Gurkha officers and NCOs would attend courses of instruction in England, mainly at the School of Infantry at Warminster and
the School of Piping in Edinburgh. As for recruits, there were now
just two depots in Nepal, Lehra, which covered the Sixth, and
Jalapahar. Here the recruit was issued with his "transit mufti"
of trousers, white shirt and Nepali hat for his railway journey
through India to Calcutta, where he stayed at the Transit Camp at
Barrackpore until a ship arrived to take him and returning leave men
to Malaya. Once he disembarked at Penang or Singapore, he then
went by rail to one of the four Regimental Training Wings. That for
the Sixth was, of course, at Sungei Patani. However, in 1951 the four
wings were to be concentrated at Sungei Patani to form the Brigade
of Gurkhas Recruit Training Centre (BGRTC), which had four
companies, one for each regiment. It also took over the Brigade Boys'
Company, which had been formed at Sungei Patani in 1949, and
administered by the 1/6th. Shortly afterwards the BGRTC became
known simply as The Depot Brigade of Gurkhas. 1951 also saw the
introduction of a new Regimental Cap Badge, with the kukris
looking more like the genuine article.

Much of the credit for setting up the Brigade of Gurkhas on such a firm foundation must go to General Charles Boucher, the first Major-General Brigade of Gurkhas. He was also GOC Malaya District and, as such, bore much of the burden of the conduct of operations during those early days of the Emergency. All who served under him will remember him for his courage, determination and cheerfulness in the face of great difficulties. It was a tragedy that ill health brought about his early retirement in 1950 and led to his death a year later.

The 1/6th remained at Bahau until September, 1951. In April Philip Townsend was relieved by Lt Col Walter Walker DSO OBE. Philip Townsend had had a hard but very successful tour in command, as the awards of the DSO and OBE to him signified, and the Battalion's efficiency owed much to him. His successor was not a 6th Gurkha, but had already established a reputation as a tough and resourceful jungle fighter, especially when commanding the 4/8th in Burma and latterly with Ferret Force. His first task was to get to know the Battalion. To this end he spent ten days at each company base and was encouraged by a tough little battle fought by C Company in June. A patrol of eighteen under Sgt Bhaktabahadur Thapa DCM was ambushed in the Jelebu area by a terrorist group at least twice the size. Two riflemen were killed in the initial exchange of fire, but undeterred Bhaktabahadur immediately launched an attack in spite of the thick jungle and the fact that the enemy were on a high ridge. The fighting went on all day until the terrorists drew off, with one definitely killed. Bhaktabahadur was awarded a bar to his DCM, and Lance Cpl Rupsing Pun received an MM for twice rescuing wounded comrades under fire. Nevertheless, actions like this were few and far between, and Walter Walker felt that the dull routine of much of the operations might well in time blunt the Battalion's edge. He was especially concerned over the Burma veterans, who had never really had the chance to recover from the war years. Thus, when the Battalion moved back to Sungei Patani for its annual retraining, he instituted a rigorous period of training at the expense of rest and recuperation. At the end of 1951, with retraining completed, the Battalion deployed to the swampy west coast of Johore.

The 2/6th continued in the Kluang area throughout 1951, and indeed stayed there until April, 1953. During their three years there they accounted for some eighty terrorists, but the man-hours to achieve these "kills" ran into many thousands, and the main problem was maintaining alertness. Nevertheless, although they did not know it at the time, the Security Forces did make considerable

progress during 1951, and the main evidence for this was a major
policy decision made by the Communist leadership in October.
Recognizing the ability of the Security Forces to operate in the
jungle, the stiffening resolve of the civil community and the fact that
terror tactics were alienating those upon whom they depended for
support, they decided to abandon pure terror attacks on innocent
civilians, as well as operations against the economy (rubber tree
slashing, attacks on tin mines etc) and the resettled villages, and to
concentrate on subversion, with merely the occasional violent act to
make their presence felt. It would be some time before this new
policy was really put into effect, but it was a major milestone.
However, for the authorities autumn 1951 was a black time. On 6
October Sir Henry Gurney, the High Commissioner, was ambushed
and killed on the Kuala Lumpur–Fraser's Hill road. The Federation
went into mourning, and, it being Dashera, British officers did not
attend Kalratri. Then, a month later, the highest weekly number of
casualties to the Security Forces was recorded, and, finally, in
December General Briggs retired, to die shortly after he returned
home.

For the Regiment, 1951 had also witnessed another sad event, the
death of Field Marshal The Lord Birdwood in May, after twenty
devoted years of service as Colonel of the Regiment. In his place,
however, came General Sir John Harding, who knew both Battalions
from his time in the Far East and had also had the 2nd Battalion
under his command in Italy during the war.

CHAPTER THREE

The Templer Era

Amid some controversy General Sir Gerald Templer was appointed in January, 1952, as High Commissioner in place of Gurney as well as Director of Operations. He arrived with the clear political aim of forming a united Malayan nation and quickly established a number of priorities – sound police work, the vital necessity of intelligence, keeping people informed of what was going on and gaining and maintaining their confidence.

For the Regiment, Templer's arrival did not mean any immediate change in tactics or policy at battalion level. The 1/6th were now at Muar, where the accommodation for all was very much better than it had been at Bahau. Although there was not a significant amount of terrorist activity, the Battalion, during its four months there, killed fifteen, including a number of District Committee members, captured two and wounded ten. During one short period three were killed in three contacts with a total ammunition expenditure of four rounds only! Clearly Walker's intensive training was paying off. Yet there were still frustrations. One carefully prepared plan called for Police Special Branch fifty miles inland from Muar to warn the Battalion when a particular MRLA platoon came into its area. The area would then be saturated, with each platoon searching one map square per day, while other battalions in the brigade provided an outer cordon. It finally happened and a B Company platoon succeeded in spotting the terrorist camp. The platoon pulled back a few hundred yards and reported back by wireless, as a result of which Walter Walker deployed the other companies in an inner cordon while B Company prepared to attack. As it did so, a camp sentry warned the occupants, who slipped through a gap in the cordon. Later the reason for this was identified as a faulty map which showed a non-existent logging track. This was to have been the intercompany boundary between A and C Companies and, not finding it on the ground, they failed to link up, allowing the bandits to escape between them. Another noteworthy aspect of this period was the Battalion's introduction to tracker dogs. However, they became bored and frustrated on long patrols, and lacked the patience of the skilled Gurkha. There were, too, combat dogs, but after one had forced one of Harkasing Rai's platoons to take to the trees, they were given a wide berth.

In April it was time to move again, this time to Kuala Kangsar in Perak, where the Battalion relieved 40 Commando RM. This was a very different area, with much dense hill jungle, and had been a terrorist stronghold from the very beginning of the Emergency. Operating from camps deep in the jungle, the terrorists often ambushed the roads in the area. Indeed, shortly after the Battalion's arrival, a jeep carrying the Intelligence Officer from HQ North Malaya Sub-District was ambushed and he, and two Tamils with him, were killed. Nevertheless Walker believed that the answer was to go after the camps rather than to concentrate on the security of the roads as the Royal Marines had done.

One of the earliest 1/6th operations in the area began on 28 April, 1952, when a terrorist surrendered to the Security Forces, bringing with him weapons which he had stolen from his comrades while they were asleep. An *ad hoc* section of six men was quickly sent to where the terrorists were, but they were alerted and threw a grenade, which injured three riflemen, and made their escape. A subsequent search of the area revealed nothing, but the SEP (Surrendered Enemy Personnel) on further interrogation said that he could show the way to the terrorist base camp. Consequently, B Company, less one platoon, moved out under Peter Winstanley on the following night and located the camp the next afternoon (30th). While Manu Gurung, the Company 2IC, set up a cordon, Winstanley, with an assault group of ten, made his way towards the camp. Then he met a group of six terrorists coming down the track from the camp. WO2 Sherbahadur shot one, while the rest fled back towards the camp. The assault group followed, reached the edge of the camp, which appeared deserted, and took cover. Winstanley, who was behind a log, was now badly wounded in the arm by a burst of automatic fire. This was returned and two or three terrorists were wounded before making off. Come nightfall the cordon was lifted, and a further search next day revealed another bandit, who slightly wounded a Gurkha in the knee before making off.

June was a particularly lively month, but it began with tragedy.

On the 2nd, an SEP informed Angus Macdonald, commanding D Company, that his platoon had just moved into the area. Macdonald sent off two platoons to try and locate them, and on the evening of the next day, to quote the official report:

"OC D Coy had passed his evening sitrep [situation report] and closed down wireless communications, when reconnaissance patrol of 12 Platoon returned to report an occupied camp . . . Maj Macdonald decided to attack this camp at last light, with 12 Pl led by the recce group, the pl advanced. On arrival at the edge of the suspect area, OC D Coy split the pl, sending half forward on either flank to cut off the escape tracks while OC D Coy with Pl HQ attacked frontally. As this assault party moved forward, they passed the flank of the enemy sentry post without seeing it, and point blank fire was opened on them. Major Macdonald received a chest wound from which he died very shortly after the action, and a GOR [Gurkha Other Rank] was severely wounded in the jaw. The flanking parties charged the camp as fire was opened, and were just able to contact the rear few enemy, killing one, wounding another. 1 rifle, 1 shotgun, ammunition and a pack were recovered."

Thus died a very gallant officer, who had won the MC in Burma with the 4/6th, and a bar and Mention in Despatches while with the Battalion in Malaya. To this day his ashes lie buried beneath a cairn in Wester Ross, overlooking Loch Maree and the hills which he, as a Scotsman, loved so much.

But revenge was not long in coming. A week later a B Company ambush party under Lt (KGO) Girmansing Thapa killed seven, including the Political Commissar of 12 Regiment, wounded Ah Thin, the Commander of 12 Regiment, who managed to escape, leaving only one of the party unharmed. Girmansing well deserved the MC awarded to him for this. On 18 June an SEP led a party to a terrorist camp, where a useful haul of arms and ammunition was found. Every month resulted in more successes, albeit occasionally at a cost to the Battalion – July two terrorists killed, August another two, October, two killed, two captured, but at a cost of six GOR casualties. The autumn, however, did provide a problem in that in August the Leave Party left for Nepal, and the recruits, who would have to be trained up in jungle warfare, did not arrive until September. Consequently the rifle companies were left with two platoons each of twenty-four men, and the Signal, QM and MT platoons had to be pressed into service, as well as the Pipes and Drums. And so it went on until the end of the year, with the Battalion beginning to dominate the area more and more effectively.

1952 also proved an adventurous year for the 2/6th. In order to control their wide area effectively it became simpler to divide itself in half, with Ralph Griffith looking after the northern sector, while Major Maynard Pockson, the Second-in-Command, looked after the south. A frustrating and, at the time, worrying incident involved Cpl Purnabahadur's section of Desmond Houston's A Company at the beginning of May. The section's wireless broke down on the second day of their patrol, and on the third their rations ran out. Expecting a resupply drop on a road marked on the map, they tried to find the DZ (Dropping Zone), but gave up on the fifth day and decided to make for the company base, having not even found the road. Back at the base concern grew for their safety, and both A and C Companies began to search the jungle for them, helped by aircraft and helicopters. No sign of them was found, and on the eighth day HQ Malaya alerted three sections of Malay Scouts to parachute into the area. But, on that very same day, a very weary and famished section arrived back at company base. The truth was that the road did not exist (another case of inaccurate maps).

To prove how easy it could be to get lost in the jungle, Captain Vyvyan Robinson of B Company went missing for three days later in the year near Kluang. He had set up the Mortar Platoon in an ambush some two and a half hours from Kluang and unwisely decided to make his way back on his own, confident of his ability and experience. Having had a number of adventures, including spending two hours in a pig-trap pit, he eventually struck the road some ten miles from where he had started and ignominiously arrived back in Kluang by taxi! Yet the 2/6th had its successes, and notable among these was a very effective ambush under Capt(KGO) Jumparsad Gurung in June, which accounted for four terrorists, including a District Committee member known as the "Coffee King" with a 10,000 dollar price on his head, and four more wounded. A little later, C Company (Major Peter O'Bree) had a satisfactory little action when Sgt Tekbahadur's platoon attacked a terrorist gang in the Yong Peng area, killing one and wounding five, and captured the first Bren gun to be recovered in Johore.

Success was also gained in another field when the Battalion won the Millington Drake Shooting Trophy. This was competed for by all the major units in Malaya, and was based on the average score of all the company teams. The Battalion was to win it for the next two years as well, thereby completing the hat-trick, a great tribute to Ralph Griffith's emphasis on weapon training.

Towards the end of the year the 2/6th found itself using a new tactic. As part of the "food denial" campaign, which was part of the Briggs Plan for cutting the terrorists off from their indigenous support, the Battalion was redeployed in the new villages and labour lines now being built to house the squatters from the jungle fringes, the terrorists' main source of food supply in the past. Here, operating within the wire perimeters of the settlements, the companies, split up into platoons all over the area, carried out patrols and food checks. Offensive operations were still carried out on the jungle fringes, but this new concept did provide a welcome break from "jungle bashing".

1953 was a memorable year for a number of reasons. Most notable was the Coronation, and both battalions sent representative parties to attend it. The 1/6th contingent, under "Mac" McNaughtan, with Gurkha Captain Manu Gurung MBE MC and Gurkha Lieutenant Lalbahadur Thapa, and that of the 2/6th under Geoff Walsh, Gurkha Captain Sarbajit Gurung MC, and WO2 Tulbahadur Pun VC, left for England in March. They returned in July, having had a more than memorable time. Then there was Colonel John Hunt's Everest Expedition, and the 2/6th provided Cpls Sobaram Pun and Prembahadur Tamang as escorts to the porters. 1953 also marked the end of Ralph Griffith's tenure in command of the 2nd Battalion, and he was succeeded by David Powell-Jones in April.

As for operations, the 1st Battalion spent much of the year involved on Operation HUNTER. It is worth recounting in some detail as it brings out how valuable the successful suborning of terrorists could be. It began towards the end of November, 1952, when a female terrorist surrendered to the police at Lenggong. Wong Kee, a mother of two young children who were being looked after by friends, was tired of being separated from them and having to live with her husband in the jungle. What was important, however, was that her husband was Leong Cheong, a District Committee Member for Lenggong. Cyril Keel, who ran the local Special Branch, saw this as a means of capturing Leong Cheong, and Wong Kee was persuaded to lead an A Company patrol to her husband's camp. On arrival, they contacted three terrorists and killed one, but there was no trace of Leong Cheong. Wong Kee admitted that she had led the patrol to the wrong camp, but was then sent to hospital on the grounds of mental imbalance. She now began to write letters to her husband, and Keel persuaded her to put in how well she was being treated and that if he were to surrender he would be well rewarded. Then Keel wrote to Leong Cheong direct and, after giving him assurance of complete secrecy and no Security Force harassment, as well as leaving food for him, eventually met him in secret. This was the first of a number of meetings between the two, and Walker himself was present at one.

It was not, however, until May, 1953, that Leong Cheong came up with some worthwhile information. He said that he was to be visited by one Kong Pak, a senior District Committee member for Lenggong and Political Commissar of No 1 Company of the MRLA's 12th Regiment. He would be coming from east of the Perak River in order to collect two LMGs, be accompanied by six to eight men, and would return via two old terrorist camps, known as Machine Gun and Broken Leg. He would pass through the former between 20 and 23 May and the latter a day later. The Battalion's task was to intercept and capture or kill the party.

The first action was to make a detailed reconnaissance of the camps, and this was done under great secrecy by Captain R. E. W. Atkins, accompanied by two SEPs and five GORs. The patrol reported back on the 19th with detailed sketches, and Broken Leg Camp was selected for the operation. However, it was possible that Kong Pak might be there on the 21st; hence there was no time to be lost. Seven platoons had been selected, one from B Company for the assault, the remainder were to act as cordon. They set off in pouring rain at last light on the 19th, reached the camp by dawn, and, having checked that it was empty, took up position. Walker himself was commanding the operation, and the key to success was to allow the terrorists a sufficient gap to pass through to get to the camp, and then to close it quickly when the firing started. It was also crucial to know when the party actually reached the camp, but luckily there was a small patch of cover within the camp itself. In this Lance Corporal Pimbahadur Thapa with an LMG and one rifleman were positioned, and they had a vine rope running back to the assault platoon in order to alert them when the terrorists arrived.

Shortly after everyone was in position, a shot rang out, but it proved to be a rifleman firing by mistake; luckily the terrorists were too far away to hear it. Some hours later the party arrived, a group of six. Pimbahadur waited until they had taken off their packs, and then pulled the vine rope, firing his LMG at the same time. The assault platoon dashed in, but five terrorists got away into the jungle. However, they were caught by the cordon, and the result was six terrorists killed and no casualties to the Battalion. Robert Atkins was awarded the MC and Pimbahadur a DCM; there was also a Mention in Despatches for Dalbahadur Gurung, the rifleman who had been with him. General Templer himself wrote to the Commanding Officer: "It was really an absolutely first-class performance in every way."

This was by no means the end of the story. During June the jungle to the north of Lenggong was searched, but without success, although there were contacts. Also, to assist Leong Cheong's cover, air strikes and intensive patrolling of his own area were carried out, he and his group being away at the time. In August Leong Cheong came up with some more information. Some terrorists from his area were being sent north by a particular route and it was decided to ambush it. This involved a long approach march and two platoons were used, one acting as a ration carrying party for the other, which was to carry out the ambush. They arrived in position on the 14th, but the terrorists did not appear until three days later. Then Lt (QGO)* Dalbahadur Gurung DCM and his platoon of C Company carried out a textbook operation. Waiting until all six were in the killing zone, Dalbahadur shot the fourth in line, at which point the remainder of the firing party opened up, killing three more. The other two got away behind and above the ambush position, where, instead of fleeing, they opened fire. Rfn Gagbahadur Pun soon put a stop to this by charging their position and despatching them both. Again, the Battalion had had a 100% success without loss, and Dalbahadur and Gagbahadur received the MC and MM respectively.

In October, as a result of further information from Leong Cheong, another operation was organized, this time to net the Commander 12 Regiment himself, in a camp in the thick jungle north-west of Lenggong. Because the camp was so deep in the jungle it was decided to fly the force in by helicopter. Prior to this, a reconnaissance was carried out by Auster light aircraft, but this scared the Commander away and, when the force did eventually arrive, the bag was but one live chicken. Nevertheless, in a subsidiary ambush, a D Company platoon killed a terrorist who was carrying a freshly killed deer. He was identified as the Commander of 12 Regiment's bodyguard by the six fingers on one of his hands. The platoon enjoyed the venison.

By now the Battalion was redeploying in Ipoh, which was wel-

* With the accession to the throne of Her Majesty Queen Elizabeth II in February, 1952, King's Gurkha Officers automatically became Queen's Gurkha Officers.

comed by all, since the families had been moved down there from
Sungei Patani the year before. However, C Company under Charles
Carroll remained in North Perak for a little longer to play out the
final phase of Op HUNTER. This time the target was Shui Fatt, a
Branch Committee Member operating east of the Perak River and
mainly concerned with controlling terrorist food cultivation in that
area. After the Broken Leg ambush, although he had worked under
Leong Cheong for some time, he became suspicious of him and it
took a while to arrange a meeting between the two, which was
eventually to be on the pretext of meeting local Communist sympathizers. The latter, however, were all disguised policemen, who
quickly overcame Shui Fatt, bundling him into a Landrover in which
he was taken under Gurkha escort to Ipoh. There he was persuaded
to lead C Company to his camp. No one was there when they arrived,
but later two terrorists returned and were killed, one of whom was
Shui Fatt's wife.

This was the end of Op HUNTER for the 1/6th, but there was a
gruesome sequel. Although the local Special Branch believed that
Leong Cheong was now too much under suspicion to be of further
use and wanted to remove him from the jungle, their masters in
Kuala Lumpur, still determined to bag the Commander of 12
Regiment, believed otherwise. The end was inevitable. In January,
1954, he was seized by fellow terrorists and strangled, while the
female courier who carried the letters between him and Cyril Keel
was brutally stabbed to death. Nevertheless, for the Battalion it had
been a rewarding operation and the award of a bar to Colonel
Walker's DSO was as much in recognition of his own part as of the
Battalion's performance as a whole.

The 2/6th finally moved from the Kluang area in April, 1953, at
the same time as David Powell-Jones took over command. Although
they had accounted for some eighty terrorists during their three
years there, they were not sorry to leave, especially since their new
operational area around the small town of Segamat in the very north
of Johore had a reputation for high terrorist activity, and both the
1/2nd Gurkhas and 1st Bn Cameronians had done well there. The
Segamat police under R. T. M. Henry QPM CPM were a very
efficient body, and the Regiment would come to know Roy Henry
well, not just because he shared a bungalow with David Powell-
Jones for twenty months, thereby providing a classic example of
close police/military liaison, but also in later years when he was
Commissioner of Police in Sarawak and finally in Hong Kong.
However, the Battalion was called back to Kluang for a short time in

May to help the 1st Bn East Yorks and 2/10th Gurkhas eliminate the Johore State Committee. Some terrorists were killed, but the State Committee was not contacted, and the most memorable aspect was the large-scale use of helicopters, which saved days of "jungle bashing"*. For the remainder of the year the time was spent mainly on food denial operations and getting to know the area. The results of the hard groundwork involved in the latter would bear fruit in 1954.

On a more domestic note, the Nepal Cup proved to be a local Derby for the Regiment for the second year running. In 1952 the 1/6th had beaten the 2nd Battalion 1-0 in the final, and in 1953 they did the same, but to the tune of 4-0. It was indeed a rare pleasure for the two battalions of the Regiment to have the chance to intermingle. One note of sadness was the death of Major Dicky Barber in a car crash during his first long leave in England.

The 2/6th began 1954 operationally in spectacular style. First off the mark was a section of Desmond Houston's A Company, which crept through the jungle in pouring rain in the Bandan area, attacked a camp and killed all six occupants. Then a patrol of D Company (Major George Lorimer) contacted three terrorists east of Labis, killing them all. Finally, Vyvyan Robinson with a Support Company (Major David Rees) platoon killed two, including a local terrorist leader, and badly wounded another two in an attack on a large terrorist camp. Unfortunately these four had left the camp just before the attack but, because they might unwittingly step on the concealed attackers, thereby compromising the whole operation, fire was opened on them. The remainder escaped, although in the follow-up action two more were killed. This brought a total bag of fifteen accounted for during the first week of January, a record which General Templer said would be hard to beat. In all the Battalion killed or captured fifty-five terrorists in 1954, a number only surpassed that year by the Fiji Battalion which had many more contacts. It showed that, like the 1/6th in 1953, the 2/6th had become a first-class fighting unit, alert, well versed in jungle lore and exhibiting high standards of marksmanship. It was fitting that it should reach this peak in the fiftieth anniversary year of its raising, the celebrations of which were marked by a number of parties and presentations at the end of the year.

The 1/6th's time in Ipoh was mainly involved in a series of large scale operations, which lacked the interest and excitement of Op

* An exception to the general character of operations was Jimmy Lys's ambush of October, 1953, in the Batu Gajah area. See Annexure A on page 114.

HUNTER, where they had been able to plan their own operations and have the excitement of working closely with Special Branch. Now they were merely one unit among many with little control over their destiny. The first, Operation INLAND, was a food denial operation in Southern Perak and took place from November, 1953, to July, 1954. It was monotonous and frustrating in the main, with successes in terms of 'kills' few and far between. One such took place in the interregnum between Walker's departure and A. E. C. Bredin's arrival, when Pat Patterson was temporarily in command. Special Branch had established that a party of terrorists was regularly collecting food outside the perimeter of a certain village. Lying up on the high ground above the village during the day to ensure that nothing suspicious was taking place, sometimes as many as twenty would move down after dark to collect the food. Nothing could be done from within the village, as Communist sympathizers would have alerted the terrorists. Thus, Patterson decided to ambush the party on its return journey. A Company were to carry out the ambush, with the remainder acting as cordon. Dry rehearsals were carried out, and on the night in question the ambush was sprung. Unfortunately the party was only four strong, two of whom were killed by A Company and the other two by the cordon.

Then, over July, August and September, came Operation TERMITE. This was the largest joint air/land operation mounted in Malaya up to this time, with five major units involved, together with three squadrons of 22 SAS, police and volunteer units. The aim was to eliminate a large group of terrorists in the deep jungle east of Ipoh, and to gain the confidence of the aborigines in the area. Preceded by Lincoln bomber attacks on known terrorist localities, the SAS dropped in by parachute. The Battalion was to deploy on the edge of the jungle and then, along with other units, move slowly in. Apart from two kills in the first few days, there were few contacts, and what stood out most to those who took part was the rugged nature of the country. C Company even reached the top of Gunong Korbu, the second highest mountain in Malaya, and the contrast between the hot, humid lowland jungle and the biting cold of the slopes of the *gunongs*, was very noticeable, especially at night. The daily tot of rum became a necessity. The overall view was that the operation was only partially successful in that it caused the enemy some disruption. Nevertheless, in spite of views from some quarters that these large-scale operations merely alerted their targets before they could be fully launched, the 1/6th was involved in two more towards the end of 1954 and on into 1955 – Operations SHARK and LADY. Both, like TERMITE, involved long and tiring periods in the jungle, with little in the way of excitement. LADY, in particular, took place in very unpleasant country – the flat, marshy and insect-ridden area around Bikam and the Slim River. SHARK also brought tragedy when on 3 July 1955 Captain Geoffrey Hart was killed leading a platoon of his C Company in an attack on a terrorist camp of eighty. A wartime 5th Gurkha, who had returned to the Army in 1953, he had made a reputation both for his thought for others and as a tireless terrorist hunter.

Before leaving this particular period of the Regiment's time in Malaya during the Emergency, here is an account of a typical 1/6th jungle patrol, written by Brunny Short, who was Second-in-Command to Speedy Bredin during the latter part of the Battalion's time at Ipoh. It reports accurately, and with humour, one among thousands of typical routine patrols carried out by both Battalions during the long years of the Emergency.

"Most jungle operations seem to start off on the MT Park at 2 or 3 in the morning. This is because the way into the jungle invariably takes you through rubber estates, vegetable gardens, or tin mining land, and for security reasons you like to get through this fringe before the labourers are up and about. There is therefore a preliminary drive, generally in an armoured 3-tonner with only the minimum sized slits affording front visibility, through the hot airless night to the debussing point.

Here the Platoon Commander takes charge, there are whispered commands, magazines are placed on the Bren guns, and the leading section disappears some unidentifiable distance off the road to shake out for the move. You utilise this pause to struggle into your pack, probably with the help of the orderly. Knowing what it contains, you may experience a brief feeling of surprise that it isn't heavier ('Why, it's not so bad after all'). This is an opinion you will have cause to revise before the day is much older.

And so you start off, probably moving with Platoon HQ, your orderly trotting along a yard or two behind you. Weapons are carried differently from the old days – no longer balanced on the shoulder with the muzzle to the rear, as in Waziristan. The rifle, Bren, carbine or whatever it is is usually hitched under the arm, muzzle for'ard, the sling over the shoulder, and the whole steadied by the hand. It is quicker to get into action when carried in this way, and more manoeuvrable in thick jungle. You may decide at some stage to try slinging the rifle in the ordinary way, but after a

brief and unproductive period in which the foresight gets entang-
led with every vine and creeper within range, you abandon the
attempt and revert to the men's system.

There is usually a pause on the edge of the jungle, whilst it gets
light enough to see by. You can now observe what your compan-
ions are dressed in – you yourself also, for that matter – and it is not
elegant. Basically there is little difference from the Burma kit, and
the colour is the same. There is, of course, the jungle hat, a
deerstalker affair whose major tactical advantage is its complete
shapelessness – it presents no hard outline in the jungle, or indeed
any visible outline whatsoever. It stays on by gravity, and sartor-
ially speaking the only polite thing to do with a jungle hat is to look
away. Then there are the jungle boots, rubber-soled and with
canvas uppers, laced on the hook-and-eye principle, stretching
half way up the calves. They are very useful when walking along
the 2-foot iron pipes of the Kinta Intake Water Catchment area,
which are met with from time to time in the jungle, but lack grip on
a wet hillside. The final note of raffishness is added by 'Rags Sweat
J. G. [Jungle Green]', worn round the neck, a dyed version of the
domestic dishcloth, and possessing similar properties.

Now you start moving through the jungle, and at this time of the
morning it is not uncomfortable. The sun has not yet had time to
make the ground surface the steam-bath it becomes later in the
day. If you are lucky, the way will be through primary jungle,
where the visibility is 30 yards or more, and the going reasonably
good even without a track. Most of the trees are surprisingly small
– a foot or so in diameter – though the occasional mastodon will
rise to 200 feet or more. The foliage canopy is about 30 feet from
the ground, and is penetrated by very little sunlight. Sometimes
the space between the canopy and the ground is filled by the attap
palm, making movement more difficult. Sometimes you get the
even less welcome pandanus palm, with vicious saw-toothed
edges. But on the whole movement through this sort of jungle is
reasonably fast.

Shortly after 9 a.m. you will hear mutters of '*Bhat pakaune*', and
at the first suitable stream a halt is made, protective detachments
are put out, and ration tins are excavated from the depths of packs.
There are two types of ration, Gurkha and British. The Gurkha
type provides two meals of rice and dhall, with fish or meat as an
adjunct, and tea at midday. The British provides two conventional meals for the early morning and evening, not unlike the wartime
K-Ration. For the midday 'snack' it produces a sticky and highly

unseasonable assortment of chocolate, Mars Bars and other sweets that literally turn the stomach in the noon-day heat. Most of us carry one British to three Gurkha, using the British at breakfast time only.

The '*bhat pakaune*' halt, which is a feature of the first day's operations only, is soon over. In a remarkably short time the debris is cleared away or concealed, and the platoon is once again on the move. By now the effect of the sun will have become increasingly apparent, and by 11 o'clock you will not have a square inch of dry clothing above the knees. The characteristic smell of sweat-drenched J.G. will start to assail your nostrils, and except for respites during the hours of darkness, will be with you now until the end of the operation.

After a further period of climbing, during which your pack, the inner side saturated with sweat, glues itself to your back like a monstrous, over-weighted clam, the pace seems to slow down somewhat. Presently the ominous word 'bans' is muttered from someone in front, and in no time you are in the thick of it. Bamboo everywhere, thickets quite impassable even to a cat, thin trailers with the strength and resilience of telephone wires, festooning the spaces between and worst of all the rotted broken-down stems that bar the way every few yards, just too high to be able to step over in comfort, and much too low to pass underneath except by going down on your knees. This process, constantly repeated, becomes exhausting, and you envy the men their low centre of gravity which enables them to duck under these obstructions with no more than a slight bend of the legs. After the top of your pack has been caught up for the tenth time, you lose patience, grit your teeth and shove. A slight splintering noise above indicates that the bamboo has given way, and in an instant your head and neck are overrun by red soldier ants, smarting with indignation at the destruction of their home, and by this time of the day in splendid fighting shape. With the help of your orderly and the rest of Platoon Headquarters they are eventually dislodged, but not before they have stung you in a dozen places or more. Next time you bend your knees.

Shortly before half past twelve comes the 'tea' halt, and as you thankfully wriggle out of your pack once more, the platoon signaller is already running out the aerial for the midday wireless call. During this call your platoon has to report position, not necessarily such an easy task as it sounds, particularly after a morning of hard going and limited visibility. A sort of *jirgah* now takes place between the Platoon GO, seated on his pack, the Platoon Sergeant and the Section Commanders standing round in a respectful semi-circle. If you are wise you keep out of this discussion: the Platoon Commander and the NCOs have much more practice in this kind of thing than you, and at this stage of the operation you will invariably be quite certain that you have travelled much further than is actually the case. So you let them get on with it and, in due course, a tentative answer is arrived at, subject to confirmation by the Platoon Sergeant who has meanwhile been sent off to a flank to make a visual recce.

By now your orderly, who has been boiling water on a Tommy-cooker, using solid fuel which looks like white soap, has got the tea ready. The milk in the Gurkha type ration comes in the form of a semi-solidified concentrate, the appearance and texture of blanc-mange, from which the orderly cuts slices as required. These he throws into the tea, where they tend to melt rather slowly. As the undissolved residue presents a somewhat unexpected appearance when you suddenly confront it in the bottom of your mess tin, at a range of about an inch and a half, when finishing off your tea, you may prefer him to use the milk from your BT ration. This works on the toothpaste principle, and though it does not look attractive whilst being added to the tea, it does behave more predictably once in the liquid.

Milap has now been established with Bn HQ and the signaller is about to pass details of the midday position when you become aware of a slight commotion. The Platoon Sergeant has returned from his recce, and it seems that through a gap in the trees he has spotted a hill that shouldn't be there. Bn HQ is abruptly told to 'wait out', the *jirgah* reassembles, and the whole business starts again.

By 1.15, if you are lucky, you are on the move once more, with about a pint and a half of tea inside you and a mental note to keep a closer watch on the map than you have hitherto.

For ease of navigation the Platoon Commander may decide to move along a stream for a bit. As a relief from continual hill-climbing you approve of this idea, until you actually get down to the stream bed. Here you find the atmosphere incredibly dank and airless, the hills close in on you from either side, shutting out the daylight, and a whole variety of fleshy, dark green aspidistra-like plants and creepers, stimulated by the waterlogged earth, flourish abundantly.

Having crossed and recrossed the stream for the twentieth time, you are profoundly relieved when the Platoon Commander decides to make for the hills once again. By this time, however, the strain is beginning to tell somewhat. The pack tends to throw you off balance, your knees lack co-ordination, and any grip on the hillside which your jungle boots ever afforded has long since disappeared. On reaching the crest of the hill, you have to take a firm line with the platoon and order a five-minute halt, a possibility which would not otherwise have occurred to them.

The last trick that fate has in store for you, before you stop for the night, is a *ladang* to move through. This is an old aboriginal cultivation, abandoned perhaps six or seven years ago, and extending for anything up to half a mile. Since sun and air have had uninterrupted access to it for a number of years, the secondary vegetation is considerable. By far the worst is a trailing fern which the men call *chungya*. It occurs also in Nepal where its only use, according to the troops, is to make bedding for new-born piglets. In the writer's opinion the pigs get a raw deal. The plant forms a solid wall nine or ten feet high, and as deep as you care to go. The task of forcing a way through it devolves on the unfortunate leading section, and progress is exceedingly slow. Needless to say the passage-way, when finally made, is a good deal too small for you yourself to negotiate in comfort, and your verbal observations after disentangling your rifle and pack for the umpteenth time are generally worth listening to. Finally, as you emerge into the open, a hanging creeper dexterously whips your jungle hat from your head and carries it smartly two or three paces to the rear. You watch it for a moment as you have watched it on twenty-nine previous occasions this day, oscillating gently just out of reach. Your orderly retrieves it for you, and you are glad to get into camp half an hour later.

On arrival in camp, there is a good deal of activity. Two sections are probably sent off immediately on local patrol, carrying nothing but their arms. They will be back in half an hour or so, and meanwhile the remainder of the platoon set about making bivouacs. Your orderly will get busy with his kukri, and in no time a site is cleared, and a framework of branches erected to take the poncho cape overhead. On the ground goes about six inches of attap palm and over that, to your surprise, another poncho cape. As you only carry one of these, you enquire where the second one came from, and, if you are proud, order that it be restored to its rightful owner. Your orderly merely grins happily and leaves it where it is.

Close by you will be the Platoon Commander, once again studying his map, for the evening wireless call is just about due. This time he has no doubt about the position at all. He has plotted progress throughout the afternoon, and finally fixed his reference by means of a stream running just below the camp – it is so close that you can hear the sound of the water. So that is that, and the co-ordinates in due course are passed over the air with firmness and decision. Shortly afterwards one of the patrols returns to camp and reports that the water is flowing the wrong way. . . .

At dusk, when everyone is in camp, Stand-To takes place – in complete silence, and actuated only by the faintest of whistles. After Stand-Down the Platoon Commander's orderly will bring you a mug with a large tot of rum in it, and another for him. Sometimes he will even accompany this with 'cocktail eats' – curried fish and bamboo shoots, chippy alu's and so on, all of which go down exceedingly well. He is followed – or sometimes preceded – by the Platoon Medical Orderly, who gives you a tablet of Paludrine, and if you have sustained any cuts or abrasions during the day, decorates them with a purplish-brown fluid which he carries in his medical *jhola*. Finally comes the *bhat* – a whole mess-tin full, with *dal*, tinned fish or meat, mashed-up vegetables from your BT ration pack, anything that can be shovelled on. You can't normally cope with all this on the first day, but your appetite improves as the operation progresses, and by the third evening you are clearing your mess tin with the best of them.

After dark there is little to do for the men. Sentries have been posted and orders for the next day sunao'd. They therefore turn in early, though subdued mutterings and sometimes chucklings can be heard beneath bivouacs for the next hour or so. You yourself will probably have another rum with the Platoon Commander, who by this time will be well away crossing the Irrawaddy or assaulting Mandalay Hill. This goes on for a while, until eventually you both decide to call it a day.

The evening chorus of jungle-bats and tree hoppers continues for a time, and then dies away. During the night, although you get plenty of sleep, you tend to wake frequently. Mostly you turn over and are immediately asleep again, though sometimes you may be diverted for a while by the gentle snoring of the GO alongside you, or the Chinese CLO [Civil Liaison Officer] talking in his sleep.

Halfway through the night, when the moon rises and floods the jungle with an unbelievable radiance, you forget the sweat and the smells and the agonies of the day before, and can only thank your lucky stars for the troops around you, whose magnificent qualities are just the same as they were when you first got to know them – maybe many years ago – and just the same as they will always be."

As far as the domestic affairs of the Brigade of Gurkhas went, a notable event took place in March, 1954, when on the 11th, Her Majesty The Queen commanded that each year, during the summer months, two Gurkha officers should be nominated to attend her as Her Majesty's Gurkha Orderly Officers. This revived a custom of the old days of the Indian Army, which had been started by Edward VII, of appointing six (later four) Indian Army Orderly Officers, and which had ceased with the outbreak of war in 1939. It was agreed that the officers would be selected from those attending courses of instruction in the United Kingdom, and one of the first two was Gurkha Captain Lalbahadur Thapa MC of the Regiment. They were to (and still do) attend Her Majesty at a number of Royal occasions, and from that first year onwards they have been honoured by Her Majesty conferring on them the Insignia of the Royal Victorian Order, Fifth Class (Member) at the conclusion of their tour of duty.

CHAPTER FOUR

The Long Road to Victory

The year 1955 marked the main turning point of the Emergency. Since 1951 the MRLA strength had fallen from 8,000 to 3,000, terrorist incidents were down from a maximum of 500 to sixty-five per month, civilian and military casualties from eighty to twelve and one hundred to fifteen respectively, and the MCP/MRLA high command had retreated to the Thai border. Indeed, by the end of the year almost 50% of the population of the Federation was living in areas free of terrorist threat. A good flow of information resulting in accurate intelligence, food denial, the resettlement programme and the targetting of the District hierarchy had all played their part. Furthermore, the general election in July, resulting in the formation of a national government under Tunku Abdul Rahman, began to weaken Chin Peng's anti-colonial platform. Nevertheless, there was still much to be done to stamp out terrorism.

For the 1/6th, February marked the shooting of the Battalion's 200th terrorist, the honour falling to an A Company platoon. In March, while the rest of the Battalion was away on Op LADY, C Company were left with the task of checking out a long extinct volcano, Gunong Rapat. It was known to offer good concealment (Jim Hannah of Force 136 had used it several times during the war) and it was likely that terrorists were operating from it. However, water was known to be short, and it was thus deduced that any camp would be near water. Young Geoff Allsop, a recently arrived Short Service officer, was sent up in an Auster to check out sources of water and located three. Then he took Lt (QGO) Dalbahadur Gurung's 8 Platoon to examine these in more detail. On the morning of 19 March, near one of the water points, they heard talking and singing, and estimated a group of about fifteen terrorists to be there – too many for his weak platoon. He called for reinforcements and Major

Derek Organ, then commanding Support Company, took one of his own platoons and another from C Company and went to join Allsop. They planned to put in an attack on the camp, which was unusually situated at the bottom of a crater several hundred feet deep with steep jungle-covered sides. How this difficult operation was carried out is related by Speedy Bredin.

"In the climb down the 'funnel' one Bren-gunner carried his weapon by gripping the sling in his teeth! The CT sentry at the bottom was surprised, though he gave the alarm before being killed. As we learnt later, from the SEP, there were not more than seven CTs in the camp at the time, under Chin Voon's second-in-command, Kok Keung, though the DCM himself had been present only two days earlier. The CTs escaped into the thick undergrowth at the foot of what turned out to be another 'funnel' of some hundreds of feet – their 'escape hatch', so to speak. Another CT was killed when two or three of them, a sort of rearguard, opened fire from concealed positions at the foot of the escape route; and two more, including Kok Keung, were wounded. But in spite of much searching and following blood trails, there was no further contact with the remaining five CTs. Their morale, however, was high and, apart from their songs on the previous day, they shouted curses at the pursuing Gurkhas, admittedly from comparatively safe points of vantage.

In recalling this action, Derek Organ has related how Dalbahadur found that he, with two other men, could get no further in his outflanking effort up the cliff, and shooting up the retreating terrorists became ineffective. It was then that the richest abuse in several languages – Chinese, Malay, Hindustani and Gurkhali – reverberated around the heart of the gunong. In a series of parting shots both sides settled on the English language 'as she is spoke' by the British serviceman all over the world. As Derek said: "I had no idea our chaps spoke so much 'English'!"

Even though our cordon was thin enough, in all conscience, we hoped that the escaped CTs were still somewhere in the gunong. Not only did 'C' and 'D' Companies alternately patrol the feature, even using an assault boat on one occasion to cross a disused mining pool, before scaling a cliff; but every available weapon, including voice aircraft and surrender leaflets, was brought to bear to induce surrenders.

The Battalion's 3-inch mortars, machine guns, 3.5-inch rocket launchers and back-pack flame throwers (used against caves at or near ground level) all had a 'go'. So did the 25-pounders of 'O' Troop 'B' Battery, Singapore Regiment RA and the armoured cars of the 15/19th Hussars with their 2-pounder guns. All in all it was a splendid display of fire-power; and the grand finale came when, on the 25th, five Lincoln bombers dropped seventy 1,000 pound bombs on the central area of the gunong, after all troops had been withdrawn to a safe distance (though patrols followed up the bombing).

The result achieved was one surrender when a frightened little man came out of that part of the cordon manned by 3 Malay Regiment.

The SEP's information tended to confirm that some of the gang were still within the feature – though it now seemed unlikely – and we maintained the cordon, hoping.

At last on the 28th the DSBO [District Special Branch Officer] brought us news that the two CTs who had been wounded were hiding in an old Chinese temple north-east of the Kramat Pulai tin mine. This meant that they had not only broken the cordon but crossed the Sungei Raia to the east. No 7 platoon of 'C' Company was immediately despatched to the scene and the information proved correct; but these two fanatical CTs still resisted and had to be shot again before they were prepared to 'come quietly'.

It was agreed that there was no point in maintaining the cordon; so all troops were withdrawn by nightfall on the 28th and the curfew over the area lifted as from 6.30am the following morning.

So ended the Gunong Rapat operation in which, out of seven CTs contacted, 'C' Company killed two, wounded and captured two and, with assistance, caused the surrender of another – a most creditable effort in a fantastically difficult piece of country.

The following message was received from the Brigade Commander on the conclusion of the operation:

'For Commanding Officer from Bde Commander. Congratulations on successful outcome of Gunong Rapat Operations. Your determined and relentless action thoroughly deserved the satisfactory result of two killed, two captured and one surrendered'.

For his gallantry and skill in the action on the 20th March Corporal Manbahadur Gurung won an immediate MM."

Much of the first part of summer, 1955, was involved with Operation SHARK, which resulted in the sad death of Geoffrey Hart, as was related in the previous chapter. At the end of July the

Battalion concentrated at Ipoh for retraining, which this time included conventional and even nuclear warfare. September, however, saw a new development. This was the introduction of an amnesty. Templer's "hearts and minds" policy was working well and was resulting not only in greater co-operation with the Security Forces by the local population, but also in increasing numbers of surrendering terrorists. Leaflets and radio broadcasts were directed at those still in the jungle, promising them their lives and an end to their hardship. In order to lend further weight to the appeal, it was laid down that the Security Forces must call out to terrorists to surrender before opening fire. However, as the following incident shows, this had its problems.

Sgt Ramsor Rai and his platoon were on a routine patrol towards the end of the year when he and one rifleman came across a camp containing ten terrorists. He sent the rifleman back for reinforcements but, because the rest of the platoon were on patrol elsewhere, only one Bren gunner was forthcoming an hour later. Undeterred, he called on the terrorists to surrender, in accordance with the Government's amnesty orders, only to receive a hail of lead in return, before the CTs turned to escape. Nevertheless, he and his two assistants managed to kill three and would very likely have bagged the lot if they had not complied with the amnesty orders. As it was, it was later discovered that one of the surviving terrorists was no less than Chin Voon, Ipoh's "public enemy number one", which made it all the more galling. Still, Ramsor Rai had the consolation of being awarded the DCM, and in January, 1956, the amnesty was scrapped.

On 1 January, 1955, the 2/6th moved from Segamat north to Seremban in Negri Sembilan, where it was to remain for the next two years. Although the camp there was rather cramped, with two of the companies and fifty of the families, who had been moved up from Kluang, in tents, it was good for the married men, and Seremban's greater range of amenities was another compensation. After two months' retraining, one company was sent to Kuala Klawang twenty miles to the north-east, but the remainder of the Battalion continued to operate from Segamat. Their first success followed shortly after this, when Major Tony Fisher and two platoons of Support Company attacked a terrorist camp, whose leader was called Tan Fook Lung, nicknamed "Ten Foot Long", whom the Security Forces had wanted for a long time. Such was the standard of fieldcraft that the cordon lay in position all night and could even hear the bandits cleaning their teeth. The assault went in at dawn, killed six and six more surrendered. Unfortunately Tan Fook Lung got away, but was killed in a bomb attack two years later.

Nevertheless, 1955, as it was with the 1st Battalion, was a much quieter year in terms of contacts and kills – an indication of how the tide was turning. Yet keenness to get to grips with the enemy was not dampened. Two illustrations of this involve Tony Harvey and A Company. In the first, late one evening a Chinese Special Branch officer passed information to Harvey of a food pick-up for that night. He immediately deployed with the assault/ambush party, while his 2IC, Jumparsad Gurung, commanded the layback cut-off party. David Powell-Jones and his Command Group also took part, in case a major follow-up developed, operating from a scout car on the Titi Road. At dawn the assault group drew a blank, but the cut-off party killed two, including a local man who had been terrorising the villagers, and whose death very much helped the "hearts and minds" campaign in the area. Later on, at Christmastime, Tony Harvey was attending a Christmas dinner at Seremban, although his company was at Kuala Klawang, when he received hot information of terrorist movement in his area. He and Geoff Walsh immediately dashed off in a scout car, eating a leg and wing of turkey apiece, and joined the Company, which was about to deploy. Unfortunately, they were misled by their guides in the dark, and the terrorists got away, although the food at their camp was still cooking.

In March, 1956, David Powell-Jones handed over command to Peter Winstanley, who was Second-in-Command at the time. His DSO and the fact that the Battalion had eliminated well over a hundred terrorists during his three years in command were indications of his leadership and ability. In particular, P-J was to be remembered for the confidence he placed in his officers and the way in which he let them get on with their jobs with the minimum supervision and interference. Indeed, it came as no surprise to the Regiment that both he and Walter Walker should get brigades in 17th Division in late 1957.

Operationally, 1956 was even quieter, albeit just as busy as 1955. By the end of it the 2/6th was looking to eliminate its 200th terrorist, but he was proving remarkably elusive. One pleasure, however, was to have 1st Bn The Rifle Brigade operating in the area immediately north, and the affiliation was furthered by the Pipes and Drums playing at their Regimental Birthday celebrations, while the Rifle Brigade sent a party to attend Dashera.

The 1/6th spent the whole of 1956 under command of 2 Federal Infantry Brigade, much of it in the Cameron Highlands. The year started spectacularly with what General Bourne, the then Director of Operations, described as the most successfully carried out company task of real difficulty during his time in Malaya. It involved C Company, then commanded by Captain (GCO) Harkasing Rai MC IDSM MM, a redoubtable and venerated Battalion character.

It all began in November, 1955, when a large group of about seventy terrorists raided Kea Farm New Village in the Cameron Highlands. Having murdered a Home Guard, cut telegraph wires, wrecked the police wireless and tried to do the same to the police station, they made off eastwards with a number of shotguns and ammunition. A large follow-up operation was launched, but several days' rain hampered positive progress. However, from two attempted terrorist ambushes it was clear that this group intended to stay in the area, and from various pieces of information it was deduced that the main terrorist rendez-vous area was around Fort Brooke, north of the Blue Valley Tea Estate and east of Gunong Korbu. C Company were deployed to the area and on 4 January came across what clearly had been an overnight resting place for about forty terrorists. Harkasing and Charles Ley, Assistant Protector of Aborigines, who was with the Company, concluded that the bandits had five days' start. Nevertheless they set off to follow the tracks. After frequent delays, when the tracks were lost (Harkasing reckoned that this was a well-disciplined band), but then found again, they came across a hut in the jungle at midday on the 9th. While surrounding it they were fired on by a sentry, who turned out to be an aborigine, who was quickly despatched. Fire was now opened from the main camp two hundred yards beyond. They charged and killed three more terrorists, but immediate follow-up of the remainder was delayed by heavy automatic fire from the other side of a river. The camp had now to be searched and large amounts of documents and equipment were found.

The Company was also out of rations, which meant another delay. In the event it was two days before they got on the move again, and followed the tracks up to a knife-edge ridge, where it appeared that the party had split into two, one going north, which could be countered by B Company which had been deployed for such an eventuality, and one south. Harkasing therefore decided to follow the latter. As the Historical Report recounted:

"If their task had been difficult before the contact, by now it was well nigh impossible. The CTs, most of them having lost their packs, moved with the wary stealth of animals. Their skill in movement, in any case almost legendary, was now increased by the fear of pursuit. They slept in caves, leaving only the smallest traces of their stay. They moved like insects along the face of cliffs, up crevices and down waterfalls, and yet always they left some all but imperceptible sign of their passing."

But perseverance was rewarded. At dusk on the 13th one of Harkasing's recce patrols came across about fifteen terrorists in a temporary camp, 7,000 yards from the original contact. Harkasing considered, in view of the fast fading light, that immediate attack was out of the question, and ordered his men to lie where they were. An hour before dawn they began to inch their way towards the camp, but as day dawned, with the camp only half encircled, they were spotted by a sentry, who let loose a long burst from his Tommy gun, wounding two Gurkhas. The terrorists, splitting into small groups, were up and away.

Meanwhile, Speedy Bredin, appreciating the need for a stop line, had deployed A Company to the south of C, and some platoons of 2nd Bn Malay Regt on the jungle edge to the west. C Company, leaving a small party to cut an LZ for a helicopter to evacuate the wounded, chased after the terrorists, but it was A Company who got the first contact, when three terrorists ran right into them. One was killed and the other two fled (one was later discovered to have been wounded). Nothing happened on the 15th, but next day Support Company, who had also joined the fray, were fired on by a further group of three, who escaped. Also, a Malay Regt rifle group in stop positions accounted for another two. Among the terrorists killed during this operation were three important members of the Perak State Communist Party apparatus, and C Company's initial contact destroyed what was to be their permanent camp. Harkasing's determination to pursue his quarry to the bitter end was recognized by the award of a bar to his MC, and Lance Corporal Indrabahadur Gurung, who was leading scout for much of the chase, received the MM.

Unfortunately for the Battalion, the remainder of the year was not to live up to this initial excitement. From 1 May they became involved in another food denial operation, operation SHARK SOUTH in the Kuala Terla/Blue Valley and Lower Cameron Highlands areas. This was part of an overall Federal plan to establish a White Belt or terrorist-free swathe across the Federation to Malacca and the west coast, and then expand it both north and south. It was hoped that the impact would be heightened since food operations had stopped during the period of the amnesty.

Operation SHARK SOUTH was mounted by first warning the public that food denial operations were to be instituted. The main object was to prevent food, clothing, medicine etc from reaching the terrorists, and movement of this was only by escorted convoy outside the villages. Fresh food was exempt, as it would probably be bad before it got to the terrorists, and each member of the population was subjected to strict rice rationing. The Battalion's first task was mainly on gate checks at the villages, searching those who went in and out, and combating attempts to smuggle. But patrols and ambushes did not stop. Indeed, anything to disrupt the terrorist supply lines was used, from identity card checks to twenty-four-hour curfews, ambushes on recognized jungle tracks, and even artillery barrages on cultivated areas on the jungle's edge and the close jungle itself. The patrols themselves were of short duration, usually between two and forty-eight hours, and concentrated on the jungle fringes. In August the gate checks were handed over to the police, allowing the Battalion to concentrate on patrolling. Contacts were few, and, like the 2/6th with the Maria Hertogh Riots, it was a challenge to be ordered to Singapore at short notice at the end of October to cope with riots brought about by a bus drivers' strike. These had quietened down by the time the Battalion arrived, and they spent a week on static guards, escort duties, road blocks and curfews before returning to the Cameron Highlands. Here heavy rains, which had started in the middle of October, became even heavier, slowing down the tempo of operations, both for Security Forces and terrorists. They were to last through to the end of the year.

As for domestic details, the main event of the year was Speedy Bredin's handover of command to Brunny Short in June. A. E. C. Bredin, although from the Dorsetshire Regiment, had quickly grown to love the Battalion, and this was reflected in his book *Happy Warriors**, an account of his time in command, but containing much else about the Brigade of Gurkhas as a whole and their part in the Malayan Emergency.

There was also another Coronation, that of His Majesty King Mahendra Bir Bikram Shah Dev of Nepal at Kathmandu on 2 May, 1956. The Brigade of Gurkhas was represented by a contingent of

---

* Blackmore Press, 1961.

eighty-seven all ranks commanded by Desmond Houston. The
Ceremony itself was spectacular, and included a Durbar. There was
much enjoyable entertainment in the week that followed, culminating in the Earl of Scarborough, representing Queen Elizabeth,
bestowing on King Mahendra the rank of honorary General in the
British Army. In her message, which accompanied the bestowal,
the Queen spoke of "the special ties of friendship binding both
countries" and "the splendid achievements of the Gurkha Regiments . . . known throughout the world". The occasion was also the
first time that Her Majesty's Brigade of Gurkhas officially met their
opposite numbers in the Indian Army, and many old friendships
were renewed.

Also, in October, 1956, a Book of Remembrance, very generously
presented by Major Jim Whittall, was dedicated by the Bishop of
Winchester and placed in the Cathedral. Among those present at the
ceremony were Lord and Lady Slim, Major General J. G. Bruce,
President of the Regimental Association, and no less than sixteen
former commanding officers of battalions of the Regiment.

1957 brought a change of scene for both Battalions. By March the
1/6th had been on continuous operations against the terrorists for
almost nine years, their score of terrorists eliminated stood at 227,
and there is no doubt that they more than deserved a break. The
news therefore that they were to move to Hong Kong at the
beginning of April was warmly welcomed. They were also, by
chance, to hand over Suvla Barracks to the 2/6th, who were delighted to find themselves in a proper barracks for the first time since
leaving India. The one disappointment for the 1st Battalion, however, was that their camp at Tam Mei in the New Territories
consisted largely of temporary accommodation, and was regarded
very much as a comedown after Ipoh.

The situation in the Colony had changed much since the 2nd
Battalion's time there in 1949–50. Although the border with
Communist China still had to be watched, there was none of the
previous tension over possible invasion threats, and internally things
were quiet, with Hong Kong fast becoming the financial market
place of the Far East. Life there was therefore very much geared to
peacetime soldiering, with its emphasis on the training cycle, guards,
routine duties, parades and visits. There was time to concentrate on
sport, and a healthy rivalry was built up with the 1/2nd and 1/10th,
who were also stationed there. One interesting aspect of the training
was the Battalion's first opportunity to work with tanks since the
war, and a very close relationship was built up with 1st Royal Tank

'Six out of Six' – General Sir Gerald Templer congratulates an A Company 2/6th Gurkha patrol on eliminating all six occupants of a terrorist camp in January, 1954. Major D. H. Houston MC is on General Templer's left.

'Queen Elizabeth's Own' – Farewell to Field-Marshal The Lord Harding after his visit to both battalions of the Regiment at Suvla Lines, Ipoh in August, 1959, to celebrate the Regiment's new title.

The Pipes and Drums march away from Buckingham Palace after the Presentation of new Pipe Banners by Her Majesty The Queen on 27 June, 1962.

2/6th Tac HQ, Bario, Autumn, 1964.

Regiment, the Colony's one armoured regiment. The seaside camp at Tai Lam, near Castle Peak, also proved very popular with men and their families. There were, of course, changes of individuals. In November, 1958, Brunny Short handed over to Wyn Amoore, who would be the last prewar 6th Gurkha to command the 1st Battalion and had commanded the 2nd Battalion in Italy in 1944–5. Brunny Short's close connections with the Regiment would not, however, be broken. Also in 1957, farewell was said to Gurkha Major Nainasing Gurung, who departed on pension after thirty-one years' service. As Speedy Bredin said of his four years as Gurkha major, "He had been a tower of strength . . . and to three commanding officers he had been 'guide, philosopher and friend', not least to myself for over two years. The Battalion owes him much, and his enduring and rock-like figure will not easily be forgotten." In short, he was the perfect Gurkha Major.

It was now the 2/6th's turn to sample the Cameron Highlands, and, although the patrolling was very strenuous and often very wet and cold at night, it made an interesting change. The only contact of note produced a narrow escape for Robin Wilson. It was in the Gunong Rapat area, well known to the 1st Battalion, at the beginning of May, 1957, that Wilson was out with the MMG Platoon following up a report of terrorists in the area. A small party opened fire from one of the many cliffs, and, as Wilson neared the top with his assault group, he saw a terrorist pull the pin of a grenade and roll it down the slope at him as he struggled to lift himself over the top. Luckily it did not go off and the platoon ended up by killing one terrorist and badly wounding and capturing another. Robin Wilson received a Mention in Despatches for his determination.

The next event of note was Merdeka, or independence for Malaya, which was proclaimed on 31 August, 1957. It had long been the intention of the British Government to grant Malaya her independence, especially in recognition of the way in which her peoples had combined together to combat the terrorist menace. Detailed plans for the handover of power had been made during late 1956 – early 1957 and in the months preceding Merdeka Day itself the handover of the administration of the Federation had gradually taken place, so that on the Day itself Tunku Abdul Rahman, as Prime Minister, took over a fully working machine. The Battalion's part in the celebrations was small, merely the provision of the Pipes and Drums to play on the Ipoh padang, and it was fully realized that this did not mean the end of the Emergency. Two thousand terrorists still remained active and, although now enemies within their own country as opposed to "liberators" of the people from the "oppression" of colonial rule, their leaders were determined to fight on.

Throughout the latter part of 1958 the Battalion continued to engage in anti-terrorist operations, although everyone sensed that the end was in sight. The area of operations became ever smaller as more and more tracts of the country were declared "white" and contacts became fewer and fewer. Indeed, it was not until the end of June, 1958, that the 2/6th finally achieved their 200th kill and the victim happened to be the cornerstone of the local Communist organization. Then the Battalion got to the stage where it was able to stand one company down completely for eight weeks. 1959 continued in much the same way and only one kill was registered, and that at the beginning of the year.

The start of 1959 was also marked by a signal honour being conferred on the Regiment when Her Majesty announced that in future the Regiment was to be called the 6th Queen Elizabeth's Own Gurkha Rifles, and representative serving and former members of the Regiment were granted an audience with the Queen at Buckingham Palace, together with the 7th Gurkhas who had been similarly honoured with the title "Duke of Edinburgh's Own".

In April, 1959, the 1/6th left Hong Kong to return again to active service against the terrorists, this time based at Kluang. The old hands in the Battalion quickly noticed how the situation had changed since they were last in the jungle, and there was more interest in a film that the Battalion was called upon to make on jungle warfare. Gil Hickey was the adviser and a platoon of C Company and others, including Jimmy Lys playing the part of a planter, the actors. The fact that the actors were ordered to let their hair grow caused some confusion in the disciplinary sense, but nowhere near that of the Brigade Commander, who, driving through the lines in his staff car, was surprised to see a uniformed terrorist shamble out of the family lines and stand smartly to attention as he passed.

In March, 1959, Pat Patterson took over command of the 2/6th from Peter Winstanley, and was to be the last of the prewar generation of British officers to serve with the Regiment. August saw a memorable event when both battalions joined together in Suvla Lines to celebrate their new title. The Colonel of the Regiment, Field Marshal The Lord Harding, was present, which greatly added to the occasion, and there was a full programme of balls, dinners, cocktail parties, sporting fixtures and a parade. The one problem was that most of the evening entertainments seemed to take place with a 0700 hours parade rehearsal next morning.

On 31 July, 1960, the Emergency was officially declared to be over and the event was celebrated next day throughout the country, with victory parades at Kuala Lumpur, Ipoh and elsewhere, to which both Battalions sent detachments. Chin Peng, Secretary-General of the Malayan Communist Party, had slunk across the border into Thailand, taking a few hundred of his hardened fighters with him. They would continue to cause the authorities some problems for a number of years and still do. Almost 7,000 CTs had been killed, nearly 4,000 had surrendered or been captured, and a further 3,000 wounded. The Regiment's share in this was well over 400 killed, and many decorations won, but it was not without cost and the Roll of Honour for Malaya lists no less than fifty-nine names. Both Battalions had spent at least 80% of the years 1948–60 on anti-terrorist operations and this bears out the fact that, together with the Malayan Police and local military units, the backbone of the Security Forces was the Brigade of Gurkhas. Lord Wavell's far-sighted recommendation of 1946 that a Gurkha element be included in the postwar British Army had been more than vindicated.

CHAPTER FIVE

Peaceful Interlude

With the ending of the Emergency the Regiment felt that it could now look forward to a period of peace such as it had not experienced since before the War. The 1/6th had ceased operations in the early Spring, 1960, and in May had gone to Borneo for company and battalion training in a jungle free area near Kota Belud, sixty miles north-east of Jesselton. In spite of the sticky heat the time there was much enjoyed, but little did they know that they would be back in Borneo in four years' time on more serious business. Then they returned to Kluang to the usual round of peacetime soldiering, with the training emphasis being on conventional and nuclear as well as jungle warfare. The 2/6th, which could boast of being the last battalion operational on the Thai border at the end of the Emergency, remained in Ipoh until July, 1961, when it returned to Hong Kong for a second tour there, now under command of Lt-Col E. T. Horsford MBE MC, who came from the 2nd Gurkhas.

As in 1949, there were no permanent barracks for the Battalion to occupy and it was scattered about the New Territories in "penny packets", with Battalion Headquarters at Beas Stables. It was now part of 48 Gurkha Infantry Brigade and the last part of 1961 saw a major internal security exercise as well as border duties, the two directions in which any battalion serving in Hong Kong had now to look. Nevertheless, there was excitement in September when Their Majesties the King and Queen of Nepal visited. 1962 was a busier year. Firstly there was a large influx of refugees from across the border – a portent of what was to happen in later years – and, in September, the Battalion had to cope with the after-effects of Typhoon Wanda, which did much damage. There was also the more than welcome move to Gallipoli Barracks in San Wai. These had only been completed eight months before and were the most modern in the Far East. Accommodation for all the men and their families, a swimming pool, football pitches and a basket-ball pitch for every platoon and the fact that the complete Battalion was now in the same lines made it a veritable paradise. The only sadness was that time was up for many of the 1948 enlistments. They had seen the Emergency through from beginning to end but they did not have the opportunity to enjoy this new home. Neither did that fine old soldier Major (QGO) Jumprasad Gurung, who left on pension before the move after thirty-three years of service, all of them spent with the 2nd Battalion. In December the Brunei Revolt erupted and the 1/2nd were hurriedly flown there, but it was to be a further five months before the 2/6th were warned for active service once again.

Another event in 1961 which deserves mention was the commissioning of Akalsing Thapa from RMA Sandhurst. It was always intended that Gurkhas would be entitled to a Queen's Commission, and Akalsing was the first 6th Gurkha to make this intention fact. Destined for the 1/6th, he and others in the Brigade gradually replaced the Gurkha Commissioned Officers, who were allowed to waste out naturally.

1962 signified for the 1st Battalion a move that would take them further away from their sister Battalion than they had ever been, as well as to pastures entirely new. Although individual Gurkhas had been attending courses in the United Kingdom for some years past, a visit there was something only experienced by the lucky few. However, in April, 1961, the Battalion was warned that it would be moving to England to become part of the Strategic Reserve on Salisbury Plain, probably at the end of the year or early in 1962. With the end of the Emergency in Malaya, combined with the shortage of barracks in Hong Kong, there appeared to be a surplus of troops in the Far East. The ending of National Service and Defence cuts had also created the threat of a manpower shortage for the British Army at home, especially in Germany, where the 55,000-men commitment was sacrosanct under NATO agreements. The War Office therefore decided to relieve a British battalion in UK for service in BAOR, replacing it by a Gurkha battalion.

The two main problems were leave and the Gurkha families. The tour would be for two years, and all those due for long leave during that time had to complete their leave before the Battalion sailed, since no Nepal leave would be possible from Britain. As for the families, only just over fifty would be able to accompany the Battalion, because of the shortage of quarters, and they were selected on the basis of the husbands' seniority, but priority was also given to those who were midwives and teachers. The remaining families had to be returned to Nepal before the Battalion sailed. Matters were eased when it was announced that departure from Kluang would not now be until early May, 1962. Nevertheless, as far as the families were concerned, much still remained to be done and the bulk of the work fell on Phyllis Castle, the Battalion's Women's Royal Voluntary Service (WRVS) lady, and on the British wives.

This is an appropriate moment to draw attention to the WRVS and their wonderful record of service with the Gurkhas. In 1949 Miss How and Miss Nunn of the WVS, as it was before being granted the 'Royal' accolade, were appointed to the 1st and 2nd Battalions respectively. Their task was to help the families adjust to the radical change of moving from Nepal to Malaya, and to look after their well-being. Much of their concern was with the health of the wives and children, especially the babies, and they always had a Gurkha *dhai* or nurse, who was a midwife, to help them, of whom the most famous was Khemkala of the 1/6th. Sewing, handicrafts and gardening were also taught and encouraged, as well as activities for the children. Their contribution to the maintenance of good morale within the battalions was, and still is, invaluable, and they have always been held in special affection.

The families moving to England had to be prepared for a very different life from that which they were used to in the Far East. It was a question of equipping them with the necessary warm clothing, of teaching them how to cope with electric cookers and look after their British married quarters with carpets and upholstered chairs, as well as making themselves understood in English shops, which meant intensive classes in English. For the men themselves, learning English and being kitted out with warm uniforms, greatcoats and battle dress, as well as a new walking-out dress of Regimental blazer and dark trousers, all of which kept the Battalion tailors very busy, meant a further flurry of activity.

Shortly after the news had been announced, Bruce Standbridge from the 7th Gurkhas took over from Wyn Amoore, who had commanded a very happy battalion for two and a half years. On 1 August, 1961, Major-General J. A. R. Robertson succeeded Field-Marshal The Lord Harding of Petherton as Colonel of the Regiment. Although not a Gurkha himself, Lord Harding had served alongside them in the First World War and had them under command in the Second, and to this day the Regiment looks back on his years as Colonel with much gratitude and pride. His successor, of course, had commanded the 1/6th during the transfer to the British Army, and his appointment was much welcomed.

Training for the 1st Battalion's new role took up much time as a result of the need to concentrate on conventional warfare. For this the Battalion used Kongkoi in Negri Sembilan, which approximated to European terrain. There was also a reorganization of the Battalion's structure, with Support Company giving way to support platoons with two 3-inch mortars and two anti-tank guns for each rifle company, the MMG Platoon becoming the Recce Platoon, and the Pipes and Drums being made stretcher bearers and Defence Platoon respectively.

By April, 1962, the Battalion was ready and Jimmy Lys went with the advance party by air, with the Commanding Officer following shortly afterwards. The families also flew in April with Phyllis Castle, to be met in England by a barrage of reporters and photographers in what was an unusually cold spring. This left the main body, which embarked in the *Nevasa* at Singapore on 4 May, with no less than four bands to play them away, culminating in a lone piper of the 2/7th playing "Will Ye No Come Back Again" as the ship passed the promontory off Blakang Mati. The voyage took three weeks, with brief stops at Colombo, Aden and Gibraltar, and it so happened that General Jim Robertson was commanding at Aden and was able to greet the Battalion on arrival there. There was a very good relationship between the Gurkhas and the crew of the *Nevasa*, which prompted her Captain, Commodore Reggie Bond, whose last voyage it was, to say, "Gurkhas are notoriously bad sailors but wonderful shipmates", which is very likely to have been echoed in 1982 during the 1/7th's voyage to the South Atlantic and back.

At Southampton the Battalion was met by the Chief of the Imperial General Staff, General Sir Richard Hull, Brigadier Brunny Short, in whose 51 Brigade the Battalion were delighted to find themselves, Bruce Standbridge and a horde of press photographers. Then to Lucknow Barracks, Tidworth, where everything possible had been done to ensure a warm welcome. However, time was short for settling in, since there was a parade for Her Majesty's Presentation of new Pipe Banners to the Regiment and a brigade exercise to prepare for.

The former took place on 27 June at Buckingham Palace, with Her Majesty presenting the Banners to the two Pipe Majors, Sgt Toyabahadur Thapa of the 1/6th and Colour Sergeant Deobahadur Ale of the 2/6th. A large crowd watched the 1st Battalion march to the Palace from Wellington Barracks to the accompaniment of Major Iain Brebner's new Regimental March "Queen Elizabeth's Own". Then the 1/6th returned to Salisbury Plain and hard training, which culminated in Exercise WINGED COACHMAN, which took place in Northern Ireland in November. The weather was bitterly cold, but the men, as usual, stood it stoically, and one rifleman who was evacuated with frostbite of the toes said that he thought that they must be all right as he "could not feel them". It was the start of one of the coldest English winters on record, with continuous snow and ice from Boxing Day until the Spring. Much of the time was spent on snow clearance and on keeping warm, but there were outings to relieve the misery, including one when the whole battalion, with families, went to Bertram Mills Circus.

February, 1963, brought a degree of uncertainty. The Battalion was scheduled to go on a NATO exercise in Denmark, but was put on seventy-two hours' notice to move to Borneo, and then also given warning of a possible move to Singapore on internal security duties. It was then decided that one company would go to Denmark, under command of the Devon and Dorsets, and Colin Scott's D Company was chosen. As the Association Newsletter said: "It was an incongruous sight to see C and D Coys, next door to each other, packing their kit. Mosquito nets, insect repellent and jungle green were C Coy's concern, whilst D wrestled with extra blankets, heavy jersies [sic] and 'Parkas' – the meteorological office reporting 17 below freezing in Copenhagen." D Company remembered Denmark chiefly for the very warm hospitality given by the Danes, and one particular vignette was that of Gurkha Lt Gaudhan Thapa waltzing with a Danish Officer's wife to the strains of *The Blue Danube*. Another who had warm memories of this time was Gurkha Captain Manbahadur Gurung, one of the Queen's Gurkha Orderly Officers for that year, who was invited by Her Majesty to go stalking at Balmoral.

However, a welcome break from the cold English climate came at the end of April when the Battalion, less C Company, went off on six weeks' training to Aden. Here on the Yemeni border the training was reminiscent of the North-West Frontier, but the only two who had prewar experience of this were the Commanding Officer and Gurkha Major Lokbahadur Thapa. The barrenness of the terrain meant that live firing exercises could be mounted with little problem and, as a finale to their stay, one such was arranged to include supporting fire from aircraft from HMS *Ark Royal*, 105 mm howitzers, tanks and armoured cars, which provided an impressive effect until, at the end of it, two camels were seen to be nonchalantly wandering about the impact area.

Next item on the programme was to run the Bisley rifle meetings, something which would become a standard task for the Church Crookham battalion in later years. The Battalion competed and came third overall, and might have won if it had not been for the old story of someone firing on the wrong target. Captain Ian Gordon also took a team up to Scotland to compete in the annual Ben Nevis Race, where it finished second to 1st Battalion The Parachute Regiment, with Lalbahadur Pun coming third and winning a silver medal. In October came a Reunion Weekend attended by many members of the Regimental Association. At it a silver statuette was presented to Field Marshal The Lord Harding in recognition of his time as Colonel of the Regiment. One particular event which did have to be modified was Dashera, in order not to shock the public and incur the displeasure of the RSPCA.

The threat of Borneo had, however, been looming over the Battalion for most of the year, and it was eventually told that it would be bound there in January, 1964. Many friends had been made in the United Kingdom, and it had been an unforgettable experience, but the general feeling was one of relief to get back to the sun.

CHAPTER SIX

Confrontation

Merdeka provided not just a setback for Chin Peng and his depleted Communist bands, but, later, for President Sukarno of Indonesia, who since 1945 had dreamed of a "Greater Indonesia" dominating Malaya and the Philippines. Although Sukarno had done little to foster this during the 1950s, it was Tunku Abdul Rahman, Prime Minister of the Malayan Federation, who provoked him into taking positive action. On 27 May, 1961, in a speech to the Foreign Correspondents Association of South-East Asia he first propounded the idea of Malaysia.

> "Malaya today as a nation realizes that she cannot stand alone. Sooner or later Malaya must have an understanding with Britain and the peoples of the territories of Singapore, North Borneo, Brunei and Sarawak."

He then went on to advocate a plan for bringing these countries into closer political and economic co-operation. Immediate reactions from the countries concerned were favourable, and overtly Sukarno appeared not to object. However, he was able to gain some support from the Malays by use of a smokescreen.

Sukarno had had a long-running dispute with the Dutch over their last remaining possession in South-East Asia, West Irian or Dutch West Guinea. In early 1962 it appeared as though war might flare up between the two and Sukarno called for volunteers from Malaya to help him rid West Irian of its colonial overlord. The Tunku made no objection and in April the Indonesians selected some 120 from over 5,000 who had volunteered from Singapore and Malaya and they were sent to Jakarta for military training and indoctrination. Meanwhile the dispute with the Dutch was solved in August and the volunteers were now organized to overthrow the Tunku Government and replace it with a pro-Indonesian one. To this end they returned to Malaya in November and December. Meanwhile Sukarno began a radio propaganda campaign against the Malaysia concept, but took no other positive steps against it.

Fiercer resistance to the concept came from another quarter. There existed in the States of Sarawak, Brunei and North Borneo (later Sabah), which bordered on Indonesian Borneo or Kaliman-tan, as it is called, a movement called the North Kalimantan National Army (Tenteri Nasionel Kalimantan Utara, or TNKU), whose object was to bring about a Confederation of the three states outside Malaysia. Brunei, with its oil riches and autocratic but benevolent Sultan, was seen as the key, and the TNKU, under an Arab-Malay called Azahari, planned that the Sultan of Brunei should head the Confederation. To bring this about, they organized a revolt in Brunei. First the Sultan was to be captured and pro-claimed Head of the new State of Borneo; then, in order to obtain arms, the police stations would be seized, and finally the rebels would capture the Seria oilfields and use European hostages and the equipment there as bargaining counters with the British Government.

On 8 December, 1962, they struck, but failed in their initial aim of capturing the Sultan, who took refuge with local Security Forces and asked for British help. Headquarters 17th Division at Seremban in Malaya already had contingency plans for just such an event, and within the next few days 1st Bn Queen's Own Highlanders, 1/2nd Gurkhas and 42 Commando RM were deployed there, and by 17 December the revolt had been crushed with the TNKU survivors fleeing to the hinterland. On 19 December Major-General Walter Walker was appointed Commander British Forces Borneo with the task of clearing the TNKU entirely from Brunei before Malaysia became established in August, 1963. This was achieved by April, 1963.

In the meantime, Sukarno, who had been growing increasingly close to China, as well as becoming reliant on Soviet aid, began to view the formation of Malaysia as a British neo-colonial plot, designed to isolate his country. Furthermore the Indonesian Communist Party had made public its support for the Brunei rebels. With internal unrest growing, Sukarno now felt forced to take more positive action. In January, 1963, he therefore set up training camps for irregular volunteers close to the border. Three months later, on 12 April, a band of these raided the police post at Tebedu, the first of a number of like operations. The frequency of these forays grew and it soon became clear that they were being led by Indonesian regulars.

This was the situation when the 2nd Battalion was finally warned, in May, 1963, that it would be going to Borneo, and it flew there at the end of June.

Slim Horsford, with Battalion Headquarters and two companies, was placed under command of 99 Gurkha Infantry Brigade Group under Brigadier Pat Patterson, which was good news. A Company (Capt J. J. Willson) was at Lawas, with platoon detachments at Long Semado, Pensiangan and Long Pa Sia, while D (Major R. N. P. Reynolds) was away on the south-east coast of Sabah at Tawau, with one platoon at Wallace Bay on Sabatik Island. The other half of the Battalion, with a small tactical headquarters under Geoff Walsh, the Second-in-Command, was deployed in the western half of the Third Division of Sarawak under command of 3 Commando Brigade. Tac HQ and C Company (Major J. S. Keen) were at Sibu and B (Lt E. D. Powell-Jones, son of "PJ", who had commanded the 2/6th in Malaya) at Sarikie, west of Sibu with a platoon at Binatang. The main tasks were preventing and dealing with cross-border incursions, carrying out "hearts and minds" operations and also, apart from D Company, training the Border Scouts, who were drawn from the border tribes and proved most useful as guides and trackers.

For the first six weeks life was quiet, and perhaps Ralph Reynolds at Tawau had the most pleasant time. He lived with his three Gurkha Officers in a Rest House overlooking the Bay, while his company were installed in a large disused warehouse nearby. Not for nothing did he become nicknamed the "Rajah of Tawau", and his three months there were relatively uneventful.

In the second half of August Geoff Walsh's command was involved in what was recognised as a marked stepping up of the cross-border campaign.* It ended with some triumph, but also with tragedy and frustration. It began with some Ibans from the Sungei Bangkit area arriving in Song on 11 August and leaving a message that there had been a Border crossing in the Ulu Ayat area. The message was not passed on by the villagers and it was a further two days before an SAS officer, who was in Song in connection with Border Scout training, got to hear about the incident. He passed what information he had been able to glean back to the operations room in Sibu. Walsh, who was also Military Commander Third Division, considered the report confused but credible, and despatched a patrol of nine led by a Gurkha officer and equipped with a radio to Song the next day. From here they were to move up the Katibas

* See map on back endpaper.

River to Manga Bangkit and obtain more precise information. That evening Major P. F. Walter, The Parachute Regiment, who was on local leave in Song, contacted the Operations Room and said that he had been speaking with a villager from Rumah Masam who said that a party of Indonesians had captured four Ibans near Rumah Man, but one had escaped and come to his village and told the story. Walter asked that he might accompany the patrol sent to investigate this, to which Walsh agreed, telling him to strengthen the patrol with whoever he could find in Song. This he did with the SAS training detachment already there, three Border Scouts, two policemen and the Kapit District Officer, making a patrol strength of twenty-one.

The patrol set off that same night (14th), following the Katibas River and the Sungei Bankit to Rumah Masam, where they interviewed the escaped Iban. They then proceeded further up the river and came across two other Ibans who had escaped. By now it was clear that a Border crossing had definitely taken place, and, although radio communications with Sibu were poor, Walsh gathered that some fifty to sixty Indonesians were involved and ordered the patrol to establish itself at Rumah Blayong, which lay astride the enemy's likely route from the border. The patrol accordingly did this, and were firm in Blayong by midday on the 15th. Walsh was now concerned to reinforce the party and, obtaining a Belvedere helicopter, arranged for fifteen men of C Company under Lt Wallace to be deployed to the Blayong area. Hugh Wallace himself had, in fact, cut short his long leave, and had only taken over from Jack Keen the day before, Keen flying back to Singapore. Because of the uncertain radio communications, Walsh was not sure whether Walter had received the message about Wallace's arrival. In the early hours of the next morning, though, he was relieved to hear that the two parties had linked up and had taken up ambush positions in the Blayong area, having heard that the Indonesians would be due there next day. Wallace himself had deployed his party in the area of the Blayong Long House, but Walsh believed that the Indonesians would not approach Blayong without making a careful reconnaissance which might well compromise the ambush. He therefore suggested to Wallace that he adopt a more offensive posture and lay his ambush further away from the Long House. Since Wallace was in a better position to assess the situation, the final decision was, however, left to him.

During the morning of the 16th communications between Walsh and Wallace became progressively worse, and the last message which Walsh received for some time was at 1100 hours when he was told that two patrols were being sent towards Rumah Man. One, under Wallace and Walter, with a radio, went by boat up the Sungei Bangkit and then south on foot along the Sungei Ayat, while Lt (QGO) Matbarsing Gurung, without radio, led the other on foot. A small base party with the second radio remained under the District Officer of Kapit at Blayong. Major Walter takes up the story:

"At one o'clock on the 16th Lieutenant Wallace, who was the leading scout at the time, stopped and beckoned me forward. He said he had seen a man in the river ahead. I looked and saw about ten on the river bank. We decided straight away they were enemy and began to deploy. The range at this stage was almost 100 yards. We were moving into position and I was moving along the opposite bank to the enemy to get in close, when I noticed a whole party of Ibans moving above me on the steep slope. At about this moment one of the Ibans let off a completely ineffective round of shotgun fire; this alerted the enemy and the excellent target of a tightly packed group of enemy on the river bank disappeared in a flash. With the PFF [Police Field Force] Corporal I called upon the enemy to surrender. They replied by opening fire themselves. Then I took up a fire position and shot the only enemy soldier I could still see. He fell into the river. I then fired into the area where the remainder had gone to ground. I was joined a few moments later by Lieutenant Wallace who had brought the Bren gun up; the Bren gunner and the Bren gun remained in my position. By this time very heavy fire was being directed against us. I noticed one bullet nick the top of Lieutenant Wallace's hat and a moment later an Iban behind Wallace was badly wounded. I had now taken over the Bren gun and was trying to keep the enemy quiet to allow Lieutenant Wallace to move to a better position with the remainder of the Gurkhas. This he did, returning after about five minutes to tell me that he was going to try to find a way over the river to get in behind. I continued to give support with the Bren gun. He then went off and that was the last I saw or heard of him alive.

Although we did not know it at the time, we had only bumped one end of the enemy position and their main position was further upstream. The next thing that happened was that the Ibans decided it was too much of a good thing and ran shouting to where they had left their boats and disappeared, taking all our kit with them; but the Gurkha wireless operator did manage to retrieve his wireless set. We attempted to make contact on the radio but without success. I then moved my position, picking up three more Gurkhas en route who had become detached from Lieutenant Wallace, and moved on the right flank of the enemy position. It was in this area that I finally broke contact at last light. Half an hour previous to this the enemy had put in what could be described as a fire-power demonstration along 300 yards of the river accompanied by a lot of shouting. Firing died down after dark but the noise and shouting continued until well after midnight. I remained just above their main position.

At first light I made the decision to return to base to pass on the facts of the situation rather than attempt to maintain contact with the enemy as I was very low on ammunition.

The morale of the enemy appeared to be fairly high, their shooting fairly accurate at close range but later on their fire control was very poor and a vast amount of ammunition was wasted.

It took me five hours' hard marching to reach Rumah Blayong. There I contacted Major Walsh and gave him the facts of the situation. He said, 'Reinforcements are coming in. Remain in your present location until further orders'. The SAS Troop had returned and I sent them off with an Iban force to construct an LZ on a likely site at the Kedai Man. This they completed in two hours and were able to take two helicopter lifts that afternoon.

By mid-afternoon it was impossible to make radio contact with Sibu so I decided at first light I would return to the scene of the contact in an attempt to discover what had happened to Lieutenant Wallace. Later that afternoon his orderly returned saying that he had left his officer very seriously wounded at last light on the 16th. Also at this time one of the guides from the cross-country platoon to the Rumah Man returned with a message from the Platoon Commander saying 'Occupied Rumah Man 1630 on the 16th, took one enemy (Kayan) prisoner, ambushing river track, suspect Indonesians very close, require food and ammunition'. This information I was able to pass to Sibu.

At first light I left the river route with a small platoon of Gurkhas for the scene of the contact. We searched the area and found the body of Lieutenant Wallace which I returned to Blayong. The area was searched and quite a lot of enemy equipment was found; this included American-type boots, American first field dressings, American webbing equipment, jungle green peak caps and empty cartridge cases including .303 cases of '56 and '57 manufacture; also a number of posters. All of this has now been handed over to Special Branch.

On my return to the LZ at Kedai Rumah Dan, which we had now established as a forward base, the Kayan prisoner was brought in. His story was that the platoon at the Rumah Man had had an action against the enemy on the morning of the 17th, he had disappeared during this action but had decided then to surrender which he did at Blayong on the morning of the 18th. He was flown back to Song almost immediately and interrogated."

Meanwhile Matbarsing's patrol had reached Rumah Man in the late afternoon of the 16th. Walsh's post-action report:

"The Long House was deserted except for three men and a woman, who had stayed to look after the house. The patrol ambushed the approaches to the house all night, and at about 0645 hours were about to withdraw to eat a meal prepared for them by the people in the Long House when a party of enemy approached. They were led by a Kayan guide dressed in civilian clothes, who came ahead to check on the Long House. He was captured by Lt (QGO) Matbarsing. A short while later Lt (QGO) Matbarsing saw a uniformed man approaching, and he opened fire, hitting him in the legs. The wounded man crawled behind a rock in the river bed, and the patrol came under fire from enemy on the far side of the river, and to their left flank, where the enemy had deployed an LMG. The patrol was pinned down in the open ground near the Long House, and a sniping battle started which lasted until 1700 hours. The wounded man remained in hiding behind the rock all this time. The patrol was unable to move except backwards, because they were enfiladed by the LMG to their left flank. They stayed in this position sniping at the wounded man hiding behind the rock in the river bed, and at any movement they could see in the heavy cover on the opposite bank of the river. By 1700 hours Lieutenant (QGO) Matbarsing had decided that there was no point in remaining where he was and, as enemy fire had died down, he withdrew his patrol to high ground behind the Long House. An effort was made to regain contact with the patrol by helicopter recce, but nothing could be seen. The patrol commander later reported that vigorous efforts were made to attract the attention of helicopters, but to no avail. The patrol later regained contact with troops in R. Blayong on 19 August."

After Geoff Walsh had finally regained contact with Walter in the mid-morning of the 17th, he arranged for another C Company

A Company 1/6th machine gun bunker at Pang Amo in 1965.

Borneo: the ambush is set.

'Hearts and Minds', a vital aspect on the campaigns in both Malaya and Borneo. The Assault Pioneer Platoon of the 2/6th constructs a bridge at Saliliran Long House, Sabah.

Hong Kong, 1967. A joint 1/6th and Police patrol checking border crossers.

The flexibility of the Gurkha – waiting to emplane on the flight deck of HMS *Bulwark*.

The Amalgamation Parade, 14 June, 1969. Lt-Col R. N. P. Reynolds
MBE takes the Salute of A Company commanded by Lt R. W. Shoesmith.

The visit of Her Majesty The Queen to Hong Kong, May, 1975. Her
Majesty inspects the Regimental Guard of Honour commanded by Major
P. D. Pettigrew. Lt-Col C. J. Scott in rear.

platoon (No 8) to be deployed and this had been landed at
a prepared Landing Zone (LZ) south of Rumah Blayong.
Accompanying it was Dr Finlayson, Director of Medical Services
Third Division, who had volunteered to provide medical cover. A
storm now caused another breakdown in radio communications and
so Walsh flew up to Song and established a small Tac HQ there with
a more powerful No 62 set. After Walter had arranged for the
evacuation of Wallace's body steps were taken to pursue and cut off
the enemy. At 1630 hours on the 18th 8 Platoon was lifted by
helicopter south towards the border and was dropped on the Sungei
Ayat on the fringe of thick primary jungle which spread south to the
border five miles away. Almost immediately they had a contact with
a party trying to evacuate two wounded by boat, killed one and
wounded three, capturing the original two wounded. Then, moving
north along the river, they shortly came across a large enemy party in
a defensive position on a hill top. They informed Geoff Walsh who
deployed further men the other side of the enemy's position. Unfortunately darkness had now fallen and the enemy were able to slip
away during the night. More troops were deployed to the area on the
19th and Geoff Walsh remained convinced that the enemy had not
yet recrossed the border.

He believed that the enemy was lying up in preparation for
another attempt to cross the border, but, while he was beginning to
redeploy his men for what he expected to be a protracted search
operation, General Walker decided on a major reorganization. The
1/2 Gurkhas were now to take over the Third Division, and the 2/6th
were to concentrate in the Brunei-Sabah-Fifth Division area. It was
very disappointing, especially as they had gone some way towards
avenging Hugh Wallace's death, and prospects, with an estimated
thirteen to fifteen casualties already inflicted on the enemy, for
totally destroying the party were good. Geoff Walsh aptly summed it
up in a "Postscript" written to his officers:

> "During the last two weeks it was obvious that our slender assets
> were stretched to breaking point, but nevertheless we achieved a
> lot. We were not allowed to stay and finish the battle we started
> ... Military movement plans remain the supreme medium for
> dispensing frustration!"

Hugh Wallace himself had joined the Battalion in June, 1959, and
soon impressed with his efficiency and dedication to his men. As to
the actual circumstances of his death, his obituary in *The Kukri* read:

"He was killed in Sarawak whilst leading a patrol of C Company which had engaged a party of about fifty Indonesians in battle. He was first wounded in the knee while trying to outflank the enemy. Unable to walk, he took cover, directed the rest of his party to continue and remained behind with his orderly whom he later despatched to base camp to call a boat or helicopter to evacuate him. When the rescue party arrived next day they found his body. Subsequent information revealed that when approached by the enemy party, Hugh, in spite of his wound, had stood up and fought it out with them until he was killed."

The only other contact of note also occurred in August and concerned A Company. A party of eleven enemy had crossed the border into the Long Semado area, but, unlike the Third Division incident, this party showed no fight, and they were, albeit after some arduous patrolling, all rounded up. The Border Scouts helped much in this, and Colour Sergeant Dhanjang Gurung of 1 Platoon deserves particular mention for the part he played. Otherwise nothing of note occurred, and the Battalion, having been relieved by the Leicesters, moved back to Hong Kong. As for their performance during their first tour in Borneo, General Walker signalled the Commanding Officer:

"On your departure for Hong Kong I wish to offer my congratulations on your successful tour here. Your Battalion has had the good fortune to be deployed where the two main contacts in the past month took place. In both contacts skilful reading of information, rapid and decisive deployment and tenacity led to the virtual elimination of two dangerous gangs at the sad cost of a promising young officer."

While the 2/6th picked up where they had left off in Hong Kong earlier in the year, albeit with an even greater emphasis on training for Borneo should they be called on again, it was now the 1st Battalion's turn to go back on jungle operations. The advance party under Gil Hickey, who was now commanding, arrived by air at Singapore on 15 January, 1964, to be followed shortly afterwards by the main body. Two weeks were spent in a transit camp, becoming re-accustomed to the heat and being kitted out for the jungle. A and C Companies embarked in the MV *Auby*, a vessel especially built for the wide rivers of Sarawak, on a very choppy voyage, which caused much sea sickness, to their operational area in the western half of the

Third Division. Initially Battalion Headquarters was at Kapit, with John Clee's A Company at Belaga and C Company, under Henry Hayward-Surry at Nanga Ga'at. D Company (Harkasing Rai) came up later in the *Auby* and took over Long Jawi from a C Company platoon. 1/2nd Gurkhas, from whom the Battalion had taken over, had had an adventurous time, but the area was now quiet, and only Jimmy Lys with B Company, which had been detached as the reserve company to the Western Brigade in the First Division, had much of interest to report in the early weeks, accounting for six Indonesians.

Towards the end of March Gil Hickey succeeded in persuading his superiors to redeploy the 1/6th and they moved to join B Company in the First Division. It was good that the West Brigade was still commanded by Pat Patterson, and he placed the Battalion in the Lundu area at the west end of the Division. Initially, however, Henry Hayward-Surry was left at Kapit, and delighted in terming his command "1/6th GR less A, B, D and HQ Companies". C did not rejoin the Battalion until May, and with B Company in reserve at Lundu, A was based at Sematan and D in platoon detachments along the border.

There was much more activity going on in this area and life was made more interesting by the adoption of a new tactic, which became the best-kept secret of Confrontation. Although there was constant information on the build-up for infiltrations and raids from across the Border, it was a question of waiting for the enemy to show his hand – a frustrating experience, especially when the expected incursions did not materialize. It was, therefore, decided to adopt a more aggressive defence policy and to catch the enemy on his own side of the border. These operations, known collectively as Operation CLARET, mainly took the form of ambushes, and an early success was achieved by Capt (QGO) Damarbahadur Gurung's 15 Platoon of D Company.

On 17 May he laid an ambush in the border area covering a recently widened track along which Indonesians were suspected of moving fairly frequently. A problem was that there were a number of houses about and much local movement between these and the fields, and it was not surprising that one of the sentries was spotted by two women who ran off screaming. Damarbahadur therefore decided to move the ambush to cover another recently used track, but again his position was compromised by locals armed with shotguns. He then decided to pretend that his was an Indonesian party, which seemed to mollify the locals, but, to be sure, he moved his ambush yet again. Leaving his platoon sergeant and eight men to establish a base, he deployed the remainder in the ambush. The platoon sergeant had been warned to be on the alert, but had strung his men out singly in a semi-circle rather than in pairs and had not checked their fire positions. Suddenly, at mid-day on the 18th, there was a burst of firing and two men at the base position were killed, but two enemy fell as well. Damarbahadur immediately moved back to the base, but discovered merely the four bodies and the packs of the base party, which he then carried back to the platoon RV, to which the survivors of the base party had already gone. Clearly the fault lay with the Platoon Sergeant with his poor dispositions of his men and their lack of alertness, but at least some damage had been inflicted on the enemy.

However, on 21 June there occurred another setback. Two patrols, totalling forty in all from 18 and 19 Platoons, which were made up of the Pipes and Drums, decided to rest for the night at a disused platoon base at Rasau, just on the friendly side of the border. It consisted of an old house, with a defensive stockade and wire nearby. Posting four sentries, the party settled down for the night with one platoon inside the house and the other underneath it. Then, achieving complete surprise, a party of about sixty enemy attacked at about 2030 hours. With little cover available to the defenders, and their wireless antenna shot away early on, the two platoons lost five killed, including WO2 Chandrabahadur Thapa, and five wounded in the fire fight which continued until 0145 hours, when the attackers, having suffered no casualties, withdrew across the border. The mistakes were clear – failure to check the surrounding area prior to turning in for the night, the use of a position which would have been well known to the Indonesians, and the lack of cover and protection – and the lessons were quickly digested. Such a thing did not happen again.

Nevertheless, the tour ended on a high note. After two inconclusive contacts at the base at Biawak, Damarbahadur Gurung's indefatigable 15 Platoon wreaked sweet revenge for their setback in May. They laid an ambush close to the border, and on 25 July a party of nine walked into it. The platoon killed five and wounded four. It was a classic ambush, and Damarbahadur well deserved the Military Cross which he was subsequently awarded.

This was the last major contact which the Battalion had, and at the end of August it handed over to the 2/10th and began to move to Hong Kong. Indeed, it was a relief to come together again, for the Rear Party had remained in Singapore, the Gurkha families in

Malacca and the British families, along with the Regimental silver, in Tidworth. With much help given by the 2nd Battalion, Gallipoli Barracks in Fanling was taken over from them while they moved to Brunei, this time accompanied by the families on a two-year tour. For the 1/6th, apart from having the chance to pause for breath after what had been a very hectic eight months, the prospect of all being together for Dashera for the first time since 1961 was a welcome one. The elements, however, combined to try to make this difficult. Firstly, luxurious as Gallipoli Barracks was, there was nowhere that would fit the complete battalion, not to mention families as well. This did not deter the Gurkha Major, Lokbahadur Thapa, who, with timber and tarpaulins, had erected a magnificent Dashera Ghar. However, during the HQ Company *bhoj* Hurricane Dot struck. The Gurkha Major's pride and joy had to be quickly dismantled, and two companies deployed to Taipo and Shatin on disaster relief. Nevertheless, undaunted, the Gurkha Major re-erected his edifice and Kalratri, although wet, was celebrated with the usual enjoyment and enthusiasm.

Another event which took place in 1964 was the Tokyo Olympics. It was the first time that Nepal had entered, and of her six entrants two were marathon runners from Nepal itself while the remainder were boxers from the 2nd Battalion. They had trained hard in Hong Kong and Jack Keen took them to Tokyo, where the Irish team also gave them much help. Bhimbahadur Gurung was drawn against an East German, who later went on to gain a bronze medal, and did well to go the full distance. Ramparsad Gurung's fight against the Hungarian and European lightweight champion was a different matter. Ramparsad had been well primed by the Irish boxers and Billy Tingle, the Battalion's honorary coach, on the Hungarian's weak points, and was certainly giving more than he was taking, when the East German referee abruptly stopped the fight just before the end of the first round. It was, as Jack Keen said, "a disgraceful decision" which a formal protest was unable to change. Omparsad Pun had the better of his Ethiopian first round opponent, who was eventually disqualified for persistent fouling, but was laid out by a tough Tunisian in the next round. As for Namsing, the final member of the quartet, he was put against a very fast hard-hitting American and the bout was stopped. Nevertheless, in spite of the disappointments, the Olympics were a unique experience for the Nepalese team and its members certainly made their mark with the other competitors.

Later that same year, in order to cement still further the alliance between the 2/6th and the 14th/20th King's Royal Hussars which had begun at the Battle of Medicina in Italy in April, 1945, approval was given for all members of the 2/6th to wear the 14th/20th's Prussian Eagle on their right sleeve,* (the 14th/20th had worn the crossed kukris on their right sleeve for some years). Shortly after, Lt William Edge came from the 14th/20th, then in Libya, to spend two months with D Company in the jungle. 1964 was also marked by Hugh Wallace's parents presenting a magnificent shield in his memory, and this came to be competed for, as the Wallace Trophy, by the companies each year. And, on handing over as Second-in-Command, Roger Neath presented the British Officers' Mess with the last Union Jack to fly from the famous Red Fort at Delhi. Neath had been Adjutant of the 2/6th at the time and had seen it lowered on the eve of Indian Independence, as the Battalion's buglers sounded Retreat.

From September, 1964, the 2/6th were back in Borneo, this time operating in the Fourth Division of Sarawak. Battalion HQ and D Company, together with the families, were at Medicina Camp, Seria, which was rented from the Shell Oil Company. The facilities here were very good, especially for the bachelor British Officers, whose Mess was one of the Sultan of Brunei's palaces. As for the bulk of the Battalion, they were under the command of Brigadier Harry Tuzo's Central Brigade, with Battalion Tac HQ under Lt-Col Tony Harvey, who had taken over from Slim Horsford, initially at Miri and then Bario, which was also the base of B Company (Vyvyan Robinson). A Company (Tim Whorlow) was at Pa Main and C Company under Jack Keen and then Pat Robeson at Long Banga. Although Robin Wilson's D Company initially remained at Seria, the companies were to rotate through there at six-weekly intervals for rest and retraining.

The company bases had now begun to resemble First World War style trench systems, with elaborate bunkers, trenches, barbed wire and Claymore mines. They each also contained a 105 mm pack howitzer, two Vickers Medium Machine Guns, an 81 mm mortar and several General Purpose Machine Guns (GPMG). Any enemy attempting to attack them would be guaranteed a hot reception. Bario itself had a landing strip much used by the RAF and Royal Malaysian Air Force Twin Pioneers, and the two helicopters

---

* In 1966 Her Majesty also approved affiliation of the 1/6th with the 14th/20th, and with effect from 15 October, 1967, they also became entitled to wear the Prussian Eagle.

permanently based there meant that it was only a few minutes journey to the other company bases. All resupply was air dropped by Beverleys. As for the local people, the Kelabits were friendly, and good relations were quickly built up with them. Some of them were in the Border Scouts and supplied sound information on enemy movements, and they were a very good source of fresh food. They lived in long houses and many was the party in them, with local rice beer and Gurkha rum flowing in equal quantities. In this area the border ran along the crest of a high narrow ridge with steep jungle-covered slopes leading up to it on both sides. The recognized crossing points were all carefully watched and one particularly good observation post had been set up on a high point on the ridge itself. Called Bunong Boedock, this became a favourite haunt of Vyvyan Robinson. On the enemy side of the border, the ridge stretched down to a cultivated plain on which Indonesian camps could be clearly seen.

The first month or so was reasonably quiet, which gave the companies time to get to know their areas well. However, whereas the former unit, 1st Battalion The Argyll and Sutherland Highlanders, had practised a policy of waiting for the Indonesians to come to them and then inserting cut-off parties on the ridge, Tony Harvey believed in a more aggressive policy and was determined to dominate the ridge itself. It was in pursuit of this that the first contact was made on 19 October when 5 Platoon of B Company bumped an enemy party. The leading rifleman was shot in the left forearm and the Indonesians then made off back across the border leaving some half-cooked rice. During the first week in November reports were received that Pa Main and Pa Lungan were likely to be attacked in the very near future. Accordingly Robinson sent the Assault Pioneer Platoon under Lt (QGO) Tejbahadur Gurung with two MMGs to Pa Lungan, while he took two platoons along the border ridge to the most likely crossing point for an attack on Pa Lungan. Making camp near the crossing, Sgt Amarbahadur Pun was sent out with a reconnaissance patrol and reported that he had smelt smoke. Early next morning a scout group under Sgt Tulkasing Gurung came under heavy fire, and Robinson attacked the camp, using his 2-inch mortar. Unfortunately one of the bombs from this hit a tree overhead and wounded Amarbahadur, but the enemy were driven off. However, unlike the CTs of Malayan days, they withdrew unwillingly, retiring a short way and opening fire once again. Indeed it took three attacks before they broke up into small parties and vanished back across the border, leaving three dead and two wounded. Two days

later it was A Company's turn when Lt (QGO) Sobaram Pun and 1
Platoon outflanked and attacked an enemy ambush from the rear,
killing two and wounding one, and earning Sobaram a Mention in
Despatches. Clearly Tony Harvey was succeeding in his aim of
dominating the border ridge.

The next stage was to tackle the enemy on his own ground and it
was again Robinson who was involved in the first Op CLARET
operation for the Battalion at the beginning of December. From the
observation post at Bunong Boedock it had been noticed that an
Indonesian patrol frequently visited a long house a mile from the
border and it was decided to ambush it. Robinson, armed with a
rifle, and a hand-picked patrol with shotguns, whose noise would not
be so obviously military when they were discharged, crossed the
border by night, slipped down onto the plain and then, as daylight
came, went to ground under a clump of bushes near a stile which the
Indonesian patrol used. Robinson had left Lt (QGO) Birkharaj
Gurung up on Bunong Boedock to warn the party of any Indonesian
activity and no sooner were they under their bushes than Birkharaj
came on the radio to report that two Indonesians were almost at the
stile. Fire was quickly opened, but neither was hit, although as they
both fled, Robinson got one in the leg with his rifle. The party had
then to scamper quickly, pursued by a quick reaction force with
tracker dogs. Now, scrambling back up the ridge in the dark, the
tracking ability of Amarbahadur pun, recovered from his wounds
of the previous month, proved invaluable, and for this and other
actions he more than merited the Military Medal and bar which
he was awarded over the next year. Later it was gathered that the
ambush although not overtly successful, had caused great conster-
nation among the Indonesians, who believed that the Kelabits were
responsible – the shotgun ploy had paid off.

While B Company had been having this little adventure, D
Company, which had relieved C at Long Banga, had run into
trouble. 11 Platoon had established an ambush position near the
border, but also had a base as well, some thirty minutes march from
the ambush. When resupply took place, a party was sent from the
base position back to the Landing Zone at Long Banga to collect the
platoon's slice. The Indonesians obviously learnt this routine and
their Marine Commandos attacked all three groups during a resupply. Radio contact was lost, Sgt Kamabahadur Gurung and two
riflemen were killed and a number were wounded. After recovery it
was found that the bodies of the dead had been mutilated. Nevertheless, the Platoon fought back well, killing five and wounding others,

and as a result its commander, Lt (QGO) Ranbahadur Pun, won an MC and Sgt Bombahadur Gurung an MM.

Throughout the first half of 1965 Op CLARET operations continued, although the objectives were always strictly limited. These mainly involved B and C Companies. The short exchanges of individuals with the 14th/20th Hussars continued, with Captain (QGO) Amarbahadur Gurung visiting the latter in Cyprus in April, and Lt John Clifton-Bligh, Sgt D. Jones and Cpl D. Tunnicliffe spending July and August with the forward companies of the Battalion. There were, too, a large number of visitors, ranging from the Duke of Edinburgh, through the Chief of the Defence Staff, to Lord Slim's son, John, who was Second-in-Command of 22 SAS at the time.

At the end of August the Battalion launched a more ambitious type of Op CLARET scheme. Although there had been no significant Indonesian incursions since the previous December, the enemy's activities on the border ridge itself indicated that he could very easily launch a major operation. Tony Harvey believed that this could be discouraged by an attack on the airstrip at Long Bawan and the positions around it. After some persuasion he obtained clearance to do this, and Vyvyan Robinson's company was selected for the task. In order to give B Company fire support, the two 105 mm's from the Bario and Pa Main fire bases were hauled up the ridge – no easy task, as it meant enlarging the LZ to be used, flying the guns and crews in, and then manhandling the former up to an observation post on the ridge. At midday on 29 August the guns opened fire on the airstrip and the storehouses nearby. B Company dropped down into the valley and wreaked further damage. Although there were no confirmed "kills", the damage done was enough to force the Indonesians to withdraw 10,000 yards from the border. Hence the operation was entirely successful, even though it made further Op CLARET operations more difficult. There were, however, one or two such "stunts", mainly involving Vyvyan Robinson, whose tireless determination to get to grips with the enemy was recognized by the award of a Military Cross to him in December, 1965. On one occasion, running short of rations on the wrong side of the border, his Kelabit guides managed to shoot three deer with their shotguns, but the patrol was gripped with the most violent stomach pains having gorged themselves on venison and had no easy time getting back. At least it seemed that no Indonesians were operating in that particular area.

One particularly intriguing occurrence happened in September.

For the past two weeks there had been trouble from Indonesian aircraft flying over the border and strafing villages and Kelabits working in the fields. The Kelabits were naturally very worried by this, but it was difficult to catch the aircraft in the act. Then, on the 17th, one of the observation posts on the ridge noticed an enemy patrol move into Long Bawang and mark out what was obviously a Dropping Zone (DZ). After forty-five minutes a C-130 Hercules appeared and began circling the DZ. Clearly the pilot was unable to identify it, but after ten minutes the Indonesian air defence batteries opened up on the aircraft. Immediately, the starboard engines were set on fire and the paratroops inside were forced to jump in a hurry, with the Hercules making a successful forced landing at Long Bawang.

By now the Battalion had been constantly on operations for over a year and in October they handed over the area to 1st Battalion The Gordon Highlanders and concentrated at Seria for a well-deserved three months' rest. Meanwhile the 1/6th had returned to Burneo for a second tour.

Warning for this came during the Christmas 1964 festivities, indeed on Christmas Eve itself, and was very much a surprise, since the Battalion had expected to stay in Hong Kong until July, 1965. The background was that an enemy build-up had been identified opposite the First Division in Sarawak. There were estimated to be 1,000–1,500 gathered opposite the Bau District and a further battalion's worth opposite Serian. West Brigade, whose area of responsibility this was, had therefore demanded reinforcements, and hence the call for the 1/6th. Thus, during the early part of January the Battalion deployed to Sarawak, taking over the Serian area from the Royal Ulster Rifles. Gordon Herring and B Company were at Tebakang, Ted Hill with C Company at Tepoi and D Company (Capt (QGO) Damarbahdur Gurung) at Pang Amo, with two platoons at Kujang Sain. Colin Scott's A Company was at Serian as Brigade reserve.

Luckily, apart from one shallow incursion south of Tebudu, the enemy made no immediate move, which gave the Battalion time to get to know the area through constant patrolling. There was some dissatisfaction with the locations of the existing fire bases, and new ones were prepared. Miles Hunt-Davis, company officer of C Company, who spent much time at Tepoi, recalls his time there:

"The base at Tepoi was built on a hill overlooking a very small school and basketball ground with the longhouse across the river.

Most of my time seemed to be spent digging enormous bunkers, trenches and communication trenches in order to house all the Tepoi garrison underground. At one stage I had eighty civilian labourers involved in the defence work. There was always one platoon out on patrol and the two platoons took this in turn. Throughout the time we were there, there was not a single contact in the Tepoi area. It was, however, a very rewarding time to me as a young officer; one really got to know one's soldiers very well."

Nevertheless, the Battalion was quickly involved in CLARET operations and a successful early ambush was carried out in mid-February by Damarbahadur Gurung, who with two of his platoons killed five enemy in a boat, allowing the accompanying civilian, who was assumed to be innocent, to get away. Two weeks later, at the end of the month, one of his platoons assisted in the rescue of two SAS men, whose patrol had bumped an Indonesian one, with both sides taking casualties. During this time Intelligence had identified three Indonesian attempts to make major incursions across the Border, twice opposite Tepoi in January and against Pang Amo at the beginning of March. In all cases it appeared that they had found themselves still well short of the Border when daylight came.

A lucky escape for Rfn Tikbahadur Pun of 15 Platoon came on 15 April. His platoon discovered an enemy camp in the Tebudu area, but Tikbahadur, who was the leading scout, was shot at by an Armalite at ten yards range, receiving one round through his hat and a flesh wound in the leg. Unfortunately the enemy then bolted and contact was lost. However, at the end of the month and in early May, two successful cross-border operations were mounted. Gordon Herring's B Company carried out the first, having been ordered by Gil Hickey to lay an ambush on the river 2,000 yds the far side of the border in the Tebudu area. In order to achieve maximum surprise Herring selected the most difficult route across the border, which meant traversing a knife edge and coping with secondary jungle. During the three-day approach march he lost his British medical corporal and the FOO's radio operator, who injured themselves in falls. He then laid his ambush and on the fifth day of the operation his perseverance was rewarded. In his own words:

"The ambush was in position once more shortly after dawn. A civilian boat passed downstream about 0830 hours. At 1105 hours there was a burst of automatic fire from the right of the ambush.

The Bren group covering this track was slightly to the rear of the ambush and facing away from the river. The young Gurkha rifleman, who had been in position behind his Bren for just over four hours, suddenly saw the enemy approaching down the track. He lay motionless, with the butt in his shoulder, until they were not more than fifteen yards away – then he squeezed the trigger. A long raking burst and the first four Indonesians dropped dead. As the remainder reeled back and then hurled themselves off the track, the lance corporal in charge of the Bren group shot another bringing the total killed to five, and, judging from the yells of pain, a further unknown number were wounded.

During this engagement another enemy party downstream opened fire on the ambush position but their aim was high and ineffective. 5 Platoon withdrew to their base and as they did so the Platoon Commander, covered by me and my runner from the top of the river bank, searched the ambush position to ensure that nobody remained. We then started to withdraw. The remainder of the enemy force chose this moment to attack the now deserted ambush position and seeing two officers going up a small bank 30 yards in front of them, immediately opened fire with automatics. The dust spurted up between our legs and on both flanks but miraculously we were not hit. The FOO [Forward Observation Officer] then brought down previously registered DF [Defensive Fire] fire on to the old ambush position, and judging from the cries of alarm from the enemy as the first salvo crashed among them, the fire had been effective. As the rate of arty [artillery] fire increased, the enemy withdrew.

Due to the proximity of our own troops, only 105 mm howitzers were able to fire. After several salvoes on the ambush position all enemy firing ceased. The weight of small arms fire put down by the Indonesians until then had been very heavy.

5 Platoon now moved back to the 6 Platoon base on the route out and here it was discovered that two members of the platoon were missing. The FOO ordered the guns to stop firing and two recce patrols were sent out to search for them but without success. 5 Platoon and 6 Platoon then joined 7 Platoon in its lay-back position and withdrew NE towards the border along the previously selected route out. The Company then halted for the night just before last light.

Early in the morning the company crossed the border into Sarawak and by midday had reached Tebudu. On arrival it was learned that one of the missing men, a young Gurkha rifleman, had returned to base before the Company. He had got lost in the jungle as he withdrew the 75 yards from the ambush position to the platoon base. He had not been with the other member of the platoon who was also lost but said that he had not seen him since they both started to withdraw. The second missing man eventually turned up about two hours after the company had returned to Tebudu.'

The two keys to success had been silent registration of artillery targets in the ambush area, and the fact that the ambush itself had been laid with all-round defence, the siting of the Bren group to the rear of the position being crucial in saving casualties.

The second such operation was carried out by Colin Scott's company against Indonesian troops billeted in a village 3,000 yards from the border. To attack the village itself would mean civilian casualties, and it was therefore decided to lay an ambush on two tracks leading to it. The village was also by a river. Since the success of previous river ambushes, the Indonesians were now clearing the river banks with patrols before sending boats up the river, and this was another factor to be borne in mind. As with Gordon Herring's ambush, the first day produced no worthwhile target, and, indeed, only two civilians in a boat were seen. However, as Colin Scott relates:

"Next day started the same way but had a very different ending. Once again all troops were in position by 0630 hours and at about 0830 hours the same two civilians came downstream in their boat. The previous day they had looked quiet and rather sleepy but on this second trip, as they approached, they were singing, and continued to do so until they were opposite 2 Platoon's most westerly group. From there they drifted past in silence until they were exactly opposite the most easterly group, when once again they burst into song. Coincidence? Suspicious? At the time 2 Platoon chose the former alternative and did not report it, but in the light of what followed it was evident that these two were indicating the extent of 2 Platoon's ambush to another group hidden somewhere in observation.

We did not have to wait long for this group to emerge, for just after 0900 hours 1 Platoon observed an Indonesian soldier moving along the track towards their ambush. He was alert, well dressed in a camouflage smock and was moving slowly and cautiously, rifle at the ready, and searching the ground to right and left. Soon it was obvious that he was a leading scout for a larger group who could be seen following behind. So, for 1 Platoon, started that most severe test of discipline and self-control – the pulse-racing wait while the maximum number of enemy moved into the killing area. At last the leading scout came opposite the flank men and the ambush was sprung. The enemy were well spread out and there were only five men in the killing area when fire was opened. Three of these were seen to be hit, fell where they were and were presumed dead. Of the other two, one was wounded but from movement in the lallang he appeared to crawl away from the battle, while the second is known to have been unscathed as he was not fired at. Through one of those errors that can too easily happen, three of our men fired at the same enemy and no one fired at the one that got away.

The enemy quickly disclosed their good training by their very speedy reaction, which was in three parts. First there was an immediate charge by about one section towards 1 Platoon's position, but this was stopped and cost the Indonesians two more dead, while one of our men was wounded in the leg. They luckily made the mistake of charging from some distance away whereas a stealthy approach followed by a sudden charge might have been more effective. The next reaction was, fortunately for us, wild covering fire from among the trees to the East while a section tried to outflank the ambush from the north, but 1 Platoon Commander had anticipated this move in his initial siting, and when the Indonesian section had lost two men killed, the remainder withdrew. Finally from the same tree line, from a position nearer the main river, an enemy group fired random 'harassing' shots towards 2 Platoon.

All this happened in about three minutes and if the enemy's reaction had been quick so had our own. As soon as fire had been opened, a ranging smoke round was called for from the 105 mm Howitzer and, true to the very high standard of the RAA (Royal Australian Artillery), the prediction and calculations were so accurate that it landed within 100 metres of where it was wanted. The target was the track through the belt of trees along which enemy reinforcements would move or the enemy who had been contacted would withdraw. As had been anticipated, the five rounds gunfire which followed had a very sobering effect on the Indonesians. Of all the weapons in our armoury, the one they feared most was artillery. It came from an unknown location without warning and there was nothing they could do about it. We were no more enamoured of enemy mortar fire, but we could do something about it, and an enemy mortar which opened up from the South bank of the river was quickly silenced by two rounds of 5.5!

It soon became obvious that if the enemy were not completely discouraged by the gunfire, they would concentrate their attention on 1 Platoon. The Company Commander therefore ordered 2 Platoon to move North and occupy a small hill to the West of 1 Platoon. This was a plan which had been discussed previously and was a position from which 2 Platoon could either bring additional fire to bear on the enemy or from where they could easily move down to reinforce 1 Platoon. The order was acknowledged and the guns continued to fire.

After the third five-round salvo, all enemy activity ceased except for three wounded men in front of 1 Platoon's position. The main body, which at the height of the contact was estimated to be forty or fifty strong, appeared to have withdrawn and the Company Commander changed his plan. Instead of an immediate withdrawal he decided to leave 1 Platoon in ambush until 2 Platoon arrived in their new position, when the second would cover the first while they did a quick sweep to collect Indonesian weapons and possibly a prisoner, if there was one who was not too badly wounded. 1 Platoon was therefore told to evacuate their wounded man back to Company HQ and wait where they were. It was to be a long wait.

The Company Commander had estimated it would take 2 Platoon about ten minutes to concentrate from ambush and move back, and when after fifteen minutes there was no sign of them they were called on the radio. No reply. Minutes stretched to the half-hour and when after that time they had still not appeared it was obvious that something had gone seriously wrong and that to leave 1 Platoon in their present location, which, although good for an ambush, left much to be desired for defence, would be foolish and asking for trouble. They were therefore ordered to withdraw to the main base while a reinforced Company HQ continued to wait for 2 Platoon.

After waiting a further half-hour this group also moved to the main base where the Company Commander decided to wait for one more hour in the hope that 2 Platoon might turn up. There were several conflicting factors which influenced this decision. The need to evacuate the wounded man called for an early move and the desirability of this course was strengthened by the thought that the longer we stayed the better chance the Indonesians had of interfering with our withdrawal route.

Against this was a natural disinclination to abandon a platoon somewhere in enemy territory and the knowledge that if we moved without 2 Platoon, we would not only be encumbered by a wounded man who could not walk, but would also have to carry 2 Platoon's packs as well.

In the end there was no choice. After about another hour there was still no sign of 2 Platoon and the decision to move was made. By this time there was no question of using tracks although this would have been much easier for the wounded man and his porters. The enemy had had over two hours in which to outflank us and ambush the tracks, so a cross-country compass march it had to be. It was an uncomfortable unmilitary party which moved off. In the lead and in the rear were sections carrying their own packs only, but the men of the other four sections and Company HQ were either carrying two packs each or were taking it in turns to carry the casualty.

Progress was very slow. Our route took us through old ladangs with their thick secondary vegetation. It had rained and the going was very slippery and particularly difficult for those carrying the wounded man. At one stage the Company Commander asked by radio for a helicopter in which to evacuate him but this was denied until we had crossed to the Malaysian side of the frontier. It was on this same radio call that we were told that 2 Platoon had arrived back at our permanent base, news that was received with mixed feelings. We were glad they were accounted for but were suspicious of their speedy return to base when they had been told to help us. Why had they gone in the opposite direction?

One more night in the Indonesian jungle and three hours' walk the next morning saw us safely across the border, and at last we were able to call for a helicopter to evacuate the casualty. Progress was then much quicker and after 2 Platoon, who had been ordered out to meet us, relieved us of their packs, in no time at all we were back in base."

This experience illustrates a dilemma in the mounting of ambushes. If kept in place until after last light, lack of visibility would have made follow-up action difficult if the ambush had been sprung. On the other hand, to withdraw before darkness fell always presented the danger of the enemy putting in a last-light patrol and gaining information on the ambush lay-out, which was probably what had

Firefighting, 1977 – combined briefing with Police and Fire Service
representatives for C Company's Guildford Detachment.

Dashera, 1978, as celebrated by A Company at Salamanca Camp, Belize.

'Handing on the Torch' – Tulbahadur Pun VC presents two recruits with their Regimental Badges after the Passing Out Parade at the Brigade of Gurkhas Training Centre in October, 1982.

happened in this case. As for 2 Platoon, the commander had decided to pull back to base after his radio failed.

Other operational incidents during the last part of the Battalion's tour included a follow-up operation after 150 Indonesians attacked the 2 Para fire base at Plaman Mapu at the end of April. The attackers, who had several casualties, managed to get back across the border. A Company, which was roped down from helicopters into cut-off positions on the border, experienced ground fire while still airborne and artillery fire on their LZs, and armed helicopters had to be used to support them on to the ground. At the end of June D Company, which was Brigade reserve, was called out after a band of Indonesian irregulars and Chinese Communists had attacked the police station at the 18th milestone on the Kuching-Serian road. A three-day search operation failed to find the attackers, and this was followed by Operation HAMMER. The purpose here was to group the Chinese in the area into five new villages, a reminder of the Briggs Plan in Malaya, and B Company was part of the cordon.

The final brush with the enemy involved a D Company fighting patrol on 29 June. Sighting an enemy party in camouflage uniforms while on a routine patrol, they laid a quick ambush. The enemy veered off at the last moment, and the platoon opened fire killing seven and wounding several others. However, they did not follow up because it was believed that there was a larger enemy force in the area. In spite of this, it was a more than satisfactory conclusion to the tour, and it was a contented Battalion which moved back to Hong Kong in July, having accounted for a satisfactory number of enemy at a cost of only two wounded to themselves.

The 2/6th made the most of their three months' rest at the end of 1965. Retraining was carried out, but there was plenty of time for sport and other leisure pursuits. The Wallace Memorial Trophy was competed for for the first time, and D Company were the winners. Nevertheless, it was all too short, and in January, 1966, the Battalion was deployed in the Interior Residency of Sabah, taking over from 1/2nd, with the families remaining at Seria. Compared to 1964–65 the area was operationally quiet, largely because of the changing political situation. Although the withdrawal of Singapore from Malaysia in August, 1965, had appeared to play into Sukarno's hands, especially as it did cause some unrest in the border states, events in Indonesia weakened his position. On 30 September there was an attempted left-wing coup, which was put down by the Army, who then began a wholesale purge of the communists. Sukarno remained in power, but his days were numbered. At the beginning of 1966 he attempted to restore his position, sacking Nasution, Head of the Armed Forces, but by then public opinion was no longer with him. Student demonstrations against the Communists and the worsening economic situation gave the anti-Communist elements in the Armed Forces their chance once more. On 11 March Sukarno handed over power to General Suharto, retaining only his title. Next day the Communist Party was dissolved, and the new leaders slowly began to take a more conciliatory view of Malaysia.

Yet patrolling was just as intensive and the three main fire bases at Sepulot, Long Pa Sia and Pensiangan were even more elaborate than before, boasting electric light, hot water and even vegetable patches. The keeping of cats became popular, and this earned ". . . the compliments of the medical fraternity for fore-thought in keeping down the rat population. [However,] the inspecting officer might have had reservations had he seen the marmalade kitten curled for warmth on a working wireless set with a saucer of condensed milk not far from his nose." Pet monkeys, mainly gibbons, were also popular, and Paul Pettigrew (OC B Company) even had a young deer, which later unfortunately leapt into the sea and was drowned while en route for Singapore. Indeed, the only two major excitements came almost at the end of Confrontation.

At the end of July, B Company was flown by Beaver from Seria to Bakelalan to help the 1/7th in tracking down a party of terrorists under a man called Sumbi. B Company had a busy and hard three weeks in which they managed to account for eight of the incursionists, before rejoining the Battalion, and 1/7th eventually mopped up the remainder, including Sumbi, by early September. Then, in early August, Neil Anderson's D Company had the last brush. Lt (QGO) Gambahadur Gurung's platoon had set up two ambushes on the Tapadang – Long Pa Sia track. He used recce patrols to ensure that these were not outflanked, and one of these, late on the second day, discovered marks on the track across the Border, and then returned, and the Platoon set off early next morning to investigate. They had not gone far when the leading scout spotted four men. Although their clothing was almost identical to that worn by the Gurkhas, the fact that one of them was carrying a Garand rifle gave them away as enemy. The Platoon quickly downed packs as the enemy scattered, and then opened fire, although the thick jungle creeper made accurate shooting difficult. Nevertheless, they wounded one and captured another.

During this time the 1st Battalion had also returned to Borneo for its third tour, arriving in January, 1966. They took over the 2/6th's old stamping ground of the Fourth and Fifth Divisions in Sarawak, with the companies based at Bario, Pa Main, Bakelalan and Long Bawan. They soon got to know the border well, but, as with the 2/6th, there was little enemy activity. However, Gil Hickey was determined to try and tempt the Indonesians into an ambush on the Border Ridge and decided to use one of the permanent observation posts, BR X, as the bait. Withdrawing the platoon from here, together with its 105 mm pack howitzer, to a new position further along the ridge on a reverse slope, he set up a permanent ambush during daylight hours at BR X. He put up a flag, and also had claymore mines laid. Eventually, on 6 May, after the ambush had been in place since 11 April, the enemy responded to the bait, and opened fire just as the ambush party was moving into position for another long day's wait. Fire was returned, including artillery, and, as the enemy withdrew behind the crest, the claymores were detonated. Unfortunately, in spite of being greased to keep out the moisture, only two of the mines exploded. A patrol was sent to follow up, and there was a further exchange of fire before the enemy party, estimated at about 100, disappeared back into Indonesia, leaving some bloodstains behind them. Major Tony Streather, who had been seconded from the Gloucestershire Regiment, and was commanding C Company, made a cross-border reconnaissance for an attack on an Indonesian airfield, but this never came off. Just before the end of the tour, 'Mac' McNaughtan arrived to take over from Gil Hickey, whose success in Borneo had been recognized with an OBE in 1965, and who left to become the Defence Attache in Kathmandu. In June, the Battalion returned once more to Hong Kong.

On 13 August, the 2/6th received a signal from Central Brigade stating that the treaty ending Confrontation had been ratified and that all patrols were to be withdrawn. The new Indonesian régime had quickly realized that the desperate state of the economy was going to lead to further unrest and that in order to obtain urgently needed foreign loans a more conciliatory attitude was required. Hence, towards the end of May, a party of Indonesian officers flew to Kuala Lumpur on a goodwill mission. This triggered off serious negotiations between Indonesia and Malaysia, although Sukarno did his best to sabotage them and was, as a result, stripped of his title of President for Life. Finally the Treaty was signed on 11 August, with Malaysia agreeing to give the people of Sabah and Sarawak the opportunity to confirm their allegiance to Malaysia.

As in the Malayan Emergency the brunt had been borne by the Brigade of Gurkhas, and the two battalions of the 6th Gurkhas had made a significant contribution. They had had their successes and their frustrations, with long periods of inactivity punctuated with short bursts of intense action against an enemy who proved himself a skilled and resourceful jungle fighter and who, unlike the Malayan terrorist, was often prepared to stand and fight rather than just melt back into the jungle.

CHAPTER SEVEN

Amalgamation

The enormous contribution made by the Brigade of Gurkhas to the outcome of Confrontation might have been expected to secure their position as part of the Order of Battle of the British Army, but this was not to be. Indeed it was Confrontation which had frustrated a Conservative Government plan to reduce the size of the Brigade, and the 1964 Labour Government, once the end of Confrontation was in sight, saw a cutting back in strength as a necessary part of its policy of economizing on defence expenditure.

The Brigade was therefore informed that the intention was to reduce it from 14,500 to 10,000 men by 31 December, 1969. Initially, although there were several rumours in the Press, Pat Patterson, now Major-General Brigade of Gurkhas, withheld the news on a "need to know" basis until reasonable terms had been agreed. Indeed, he was faced with an unenviable task since it was bound to cause disappointment, sorrow and even anguish and not everyone could be satisfied. It is a measure of General Patterson's determination and integrity that there was no discord within the Brigade and that the eventual terms were as reasonable as they were.

General Patterson evolved four main principles that were agreed by the Ministry of Defence. Firstly, the rundown would not exceed 2,200 all ranks in any one year, since the Resettlement Centre in Nepal could not cope with a higher rate. The rate of rundown would be the same for all units, and no man would be declared redundant without a minimum of six months' warning. Finally, and this was looking to the future, a minimum of 300 recruits would be enlisted each year. However, the financial aspect was more troublesome. Initially the Government offered the compensation terms drawn up by the Conservatives in 1962 for those who had to go before their full term. A committee was set up by General Patterson to deal with this problem and, on looking at the initial proposals, they were not reassured. The proposals revealed almost total ignorance of the original Terms and Conditions of Service drawn up under the Tripartite Agreement of 1947 and took little account of Nepal's economic situation. The Committee therefore had to take up cudgels with the Treasury and General Patterson refused to make any public announcement on the rundown until the matter was settled. Eventually acceptable terms were worked out and the details were in the hands of Commanding Officers by 8 December, 1966. Three days later they made them known to the battalions.

For the Regiment the blow was a bitter one in that the rundown meant the disappearance of the 2nd Battalion and this was to be completed by June, 1969. It was, as the Colonel of the Regiment said, "a great disappointment", but he asked everyone, serving and retired, "to accept it in the right spirit". For the 2/6th, the one consolation was that it was amalgamation rather than wholesale disbandment, and the Regiment regarded it as the re-absorption of the old "Left Wing" of the Regiment, which had been separated in 1904 in order to form the 2nd Battalion. The actual work of ironing out the details of who was to go and who to stay fell very much on the two Adjutants, Mike Whitehead of the 1/6th and Evan Powell-Jones of the 2/6th. In the meantime life had to go on.

When the 1/6th arrived back in Hong Kong in July, 1966, the Colony was peaceful, but there were always the twin threats of trouble brewing on the border and internal unrest, and the Battalion had to quickly adapt back from "jungle bashing". Luckily the new Commander 48 Gurkha Infantry Brigade, Brigadier Peter Martin, was a man of great imagination, and training conducted by him always had the element of the unexpected. During one internal security exercise, he

". . . arranged a farewell Cocktail Party for the local (notional) Governor on the second evening of the Exercise, but also . . . [for] . . . the abduction of the Governor by exercise rebels within seconds of his departure from the Party.

The first that Colonel Donald and those of us attending the Party knew about this was the reappearance of the Brigadier staggering up the steps supported by two stalwarts. He was bleeding profusely from a wound in his stomach and looked as though he was in the last stages of consciousness! His acting was superb. He was obviously too far gone to be able to speak but a willing helper at the scene of the abduction managed to scribble down his last words. These were: 'The Governor has been abducted. Find him. Quickly'.

Twenty-six hours and two major redeployments later, the Governor and those of his captors who had avoided capture earlier, were picked up fifteen miles from the scene of the abduction by Philip Powell-Jones at an isolated embarkation point on the south coast near Castle Peak Hotel. This particular Governor was a cross-country enthusiast."

But life was not all work. Two social events which are still remembered were the Brigade of Gurkhas Ball held in the Peninsula Hotel and the British Officers' Mess fancy dress party with its theme of the shipwreck of SS *Gallipoli*. Sport, too, was very prominent, but, although the Battalion had won the Nepal Cup in 1964, they had to bow out in the first round in both 1965 and 1966. However, the cross-country team, under the very experienced tutelage of that old warrior Harkasing Rai, covered themselves in glory, winning both the Land Forces and Hong Kong AAA championships. One moment of sadness was the death of Khemkala, the Battalion's midwife, in November, 1966. It was calculated that during her twenty-six years in this capacity she had delivered no less than 2,600 babies! A few months later her close friend and workmate Phyllis Castle retired from being the WRVS representative, but not before her ten years of devoted service to the Battalion's families had received official recognition in the shape of an MBE. She was succeeded by Roddy Carpenter, who remained with the Regiment until 1983, and who was also made an MBE for her efforts.

The first few months of 1967 continued to pass quietly, with the Battalion being involved mainly in collective training, but at the beginning of the Summer the situation in the Colony took a turn for the worse. The cause was Chairman Mao's Cultural Revolution and the resultant upheaval and unrest spilled across the border into Hong Kong. In early May rioting broke out both on Hong Kong Island and in Kowloon, and the Battalion was deployed to the Kai Tak area in support of the Police. The latter managed to get the situation under control and the Battalion moved back to barracks, where it remained as the Colony reserve. This meant that it was on continuous stand by, from immediate to four hours notice, and all activities had to be concentrated in the lines with all married officers living in barracks. It was during these hot months of confinement that the Gallipoli swimming pool really came into its own.

In early July the trouble area switched. On the 8th, as the Police were marching up the road to relieve their frontier post at the sleepy little village of Sha Tau Kok at the very eastern end of the border, the Chinese Communist Army troops on the other side opened fire causing several Police casualties. Immediately, 48 Brigade was ordered to reinforce the border, with the 1/7th and 1/10th being deployed there, while the 1/6th acted as reserve for both the border and the Colony. This meant having equipment ready at all times for both internal security and conventional warfare, which required great flexibility. The Battalion was called upon to carry out internal security tasks first when, on the 15th, rioting broke out in Tsuen Wan. Although the Police eventually succeeded in restoring peace to the area, two companies took part in a cordon and search operation during which a number of Communist agitators were apprehended by the Police, including one shot and killed. However, incidents on the border showed that two battalions were not enough and the Battalion was ordered to take over the Ta Ku Ling sector.

The tension on the border was such that it was very difficult not to upset the susceptibilities of the locals. A classic instance of this came in August. The Commanding Officer of the 1/10th was visiting the border at Man Kam To and found himself, in the company of the Tai Po District Officer and a senior police officer, surrounded by angry villagers who seemed intent on jostling him across the border. It took some hours to calm them. As a result the Government ordered the closing of all gates in the frontier fence, and this led to a number of incidents, including an unpleasant one involving Major Bob Duncan and B Company. He was at Ta Ku Ling where the locals on the other side were angry at the closing of their gate, as it prevented them from coming across to work their fields on the Hong Kong side. So upset were they that they got through various holes in the fence and attacked Kharkaparsad Pun's platoon with stones, staves and anything else they could get their hands on. Bob Duncan, Kharkaparsad and a number of the men were badly bruised and, because the Government had laid down that there was to be no shooting "unless life was truly in danger", they were forced to take shelter in the Police Station. Mac McNaughtan soon arrived with reinforcements, and a few hours later the Hong Kong Government relented and agreed that the gate should be re-opened. Accordingly, in order to convince the crowd of his peaceful intentions, Bob Duncan took only his orderly and a small police escort to do this. He opened the gate, but was then surrounded by the crowd, who refused to let him go until he had signed a statement apologizing for the closing of the gate and for other "provocative" acts and promising "better behaviour" in the future. It took a strong force of Gurkhas and police, with much hand-to-hand fighting, to extract Bob and his party and there were a number of injuries on both sides. However, even more worrying was that, while all this was going on, a force of Chinese soldiers in more than company strength appeared on the other side of the border and occupied a series of prepared, dug-in fire positions, which dominated the area. In their efforts to extract Bob from his predicament, a police section fired rubber bullets, which landing in wet padi, produced spray very like a strike of small arms fire. The Chinese soldiers obviously thought this and let loose a long burst of automatic fire over the heads of the troops and police. As Mac McNaughtan said, "for a moment, [it] seemed to bring World War III appreciably closer".

For the next month the situation remained tense and culminated, at the end of September, in a police inspector being abducted across the border, although he managed to escape back again. The result of this was that protection of the border was made more the Army's responsibility and tours there became very much part of soldiering in Hong Kong. The fence itself was strengthened, strongpoints were constructed and, for a time, even anti-personnel mines were laid, but these proved more of a hindrance than a help and were later removed.

Sadly, at the beginning of 1968, Donald McNaughtan had to relinquish his command. He had been in and out of hospitals since the previous October with a duodenal ulcer, and eventually he was forced to go back to England to have it treated. During this time Jimmy Lys had stood in for him, but then John Whitehead came across from the 7th Gurkhas to command the Battalion up to amalgamation. Another of the ever-increasing number of departures which took place, this time later in the year, was that of Major Harkasing Rai MC and Bar, IDSM MM. Since his transfer from the 1/10th in 1952, Harke had been, with his delightful family, one of the stalwarts of the Battalion. Indeed, he was unique, and a fitting description was once provided by an American journalist, who saw him as "a square-cut hunk of carved teakwood who holds four decorations for gallantry and deserves one more for his quiet sense of humour". Otherwise the year was mainly built around duties on the border, but there were few of the excitements of 1967. There was, however, one occasion when the police were practising firing tear gas cartridges on their range some thousand yards back from the border. The wind unfortunately carried the gas over to some farmers tilling their fields just on the Hong Kong side. The latter immediately and understandably took umbrage. In no time a mob had gathered, which did extensive damage to Lo Wu Railway Station and it took two companies to evict them.* Towards the end of the year one company at a time began to go to Malaya for a period of jungle training, which provided a welcome break. Then, in early 1969, preparations for the arrival of the 2nd Battalion got under way.

The 2nd Battalion had remained in Seria until December, 1966, having handed over its company bases in Sabah to the Royal Malaysian Regiment. Perhaps the highlight of these last few months in Borneo was Dashera – the first time the Battalion had been together for this for three years. It also marked the departure of the Gurkha Major, Khusiman Gurung, who took his leave after thirty years' loyal service. Also in December, news came through of the award of the OBE to Colonel Tony Harvey, though he was no longer in command, having handed over to Roger Neath the previous May. The Battalion now returned to Kluang, so long its home during the Emergency. Everyone was concentrated there by Christmas, and those who had been there in the early fifties were pleased to see that the saplings had become fully-fledged trees, grass had replaced bare earth and proper buildings the old attap huts.

The 2/6th found itself part of 3 Commando Brigade and, not unnaturally, its first exercise in January, 1967, was an amphibious one, with the Battalion being embarked in the Assault Landing Ship *Fearless* and Commando Carrier *Bulwark*. The Royal Marines were full of praise for the men's efficiency and keenness, while the Gurkhas enjoyed the British rations and, above all, the rum issue. Much of the time was spent in being "enemy", not just to the Marines, but to other units as well. During one of these exercises there was a repetition of an earlier occurrence at Kluang towards the end of the Emergency (see page 54), when a rifleman was sent off disguised as a female rubber tapper to locate some advancing Australians. As before, discipline was stronger than all else, and the Divisional Commander driving up the road in his staff car was amazed to see a small Chinese woman spring to attention as his car passed through a village. An added interest was that the battalion could now boast its own air platoon of helicopters under Robin Adshead.

Thus passed 1967, but early the following year the 2/6th moved at short notice to Hong Kong, to help with the border and with internal security, leaving the families in Kluang. Here it was based in Kowloon, with Battalion HQ and D Company at RAF Kai Tak, C

* See also Annexure B on p. 117.

Company in Whitfield Barracks, conveniently close to the "bright lights", and A and B in the rather less comfortable Chatham Road Camp and, later, Sai Kung Camp. They had two short spells on the border, and an interesting challenge was one that faced Evan Powell-Jones. Inflammatory Communist posters were discovered pasted on a water tank in Sha Tau Kok, but to remove them during the day would be regarded as provocative by "the other side" – the tank was within stone-throwing range of the border – and, in any event, they were made of very thin paper and stuck with very strong glue. Yet, to leave them would be definite loss of face. Thus, Powell-Jones's neat solution to the problem was to paste road safety posters over them during the night. Otherwise life was quiet and is best summed up by a typical border OP report: "One CCA [Chinese Communist Army] with no raifal but bicycikle came and went." However, within the Colony much time was spent on "hearts and minds" projects including repairing an old fort as a tourist attraction, constructing paths and restoring a bridge. It was also a good opportunity to get together with the 1st Battalion in order to discuss the forthcoming amalgamation.

In June, 1968, the Battalion returned once more to Kluang, where, in spite of a continual rundown in strength, life was still one of exercises. These went on until the end of March, 1969, when the Battalion was declared non-operational so that it would have time to get ready for its final move back to Hong Kong. November, 1968, had seen the departure of Roger Neath, and Ralph Reynolds, who had been selected to command the amalgamated battalion, took his place. On 18 April, 1969, the last Medicina Parade as a separate battalion took place, with the salute being taken by General Sir Peter Hunt, and then, at the beginning of May, the move to Hong Kong began.

By 3 June the two Battalions were gathering together in Gallipoli Barracks under the command of Ralph Reynolds, and ten days later, on the 14th, came the formal Amalgamation Parade. It was a simple domestic affair, with no outside guests. First the Commanding Officer inspected the parade and made a short address. Then came the symbolic reunion of the two Battalions when the Queen's Pipe Banners were trooped, one from each flank of the parade, to meet in the centre and then together join the Pipes and Drums. An exhortation from the Gurkha Major Pahalsing Thapa, the advance in review order, march past and then the amalgamated Regiment marched off. A new era had begun.

CHAPTER EIGHT

Of Recent Memory

The 6th Gurkhas remained in Gallipoli Barracks until Autumn, 1971. Luckily both the border and Hong Kong itself remained quiet, which gave the Regiment the chance to get used to being a single battalion. Yet, it was still a sad time as more and more familiar faces vanished from the scene. These included Vyvyan Robinson, Peter O'Bree, Lawrence Pottinger, Neil Anderson, Colin Fisher, Tim Whorlow, Teddie Wilkie, Robin Adshead, Hariprasad Dewan, John Willson, Tekbahadur Subba and many others who had helped to create the history that has so far been written on these pages. Just before Amalgamation, General Jim Robertson handed over Colonel of the Regiment to General Pat Patterson, who had done so much to make the reduction of the Brigade of Gurkhas go smoothly. General Robertson's eight years in office had seen much happen, including Confrontation and Amalgamation, and it was he who obtained official recognition for the Regiment's affiliation with the 14th/20th Hussars and affiliation to the Royal Green Jackets in place of the earlier affiliation of the Rifle Brigade. But, saddest of all, was the death in 1970 of the Regiment's most distinguished soldier, Field Marshal The Viscount Slim KG GCB GCMG GCVO GBE DSO MC.

It was also a time of continuing uncertainty. The initial reduction had been aggravated by Denis Healey's 1967 White Paper with its policy of withdrawing as much as possible from East of Suez, and at one time it looked as though the Brigade of Gurkhas would be reduced to four battalions only. In the event, thanks to the Sultan of Brunei agreeing to pay for a Gurkha battalion to be permanently based in his country, the 2nd Gurkhas managed to retain both battalions, and the Gurkha Engineers, Signals and Transport Regiment were also saved. The new deployment from now on would be three battalions, along with supporting arms and services in Hong Kong, one battalion in Brunei and one at Church Crookham in England. Also, in 1971, Training Depot Brigade of Gurkhas was moved from Sungei Patani to Hong Kong, and set up by Colin Scott as the Training Centre Brigade of Gurkhas at Shek Kong.

Although Hong Kong itself may have been quiet, the endless round of border duties and guards, as well as rigorous training exercises, kept the Regiment busy. One particular pleasure at this time was to have B Squadron 14th/20th Hussars as part of 48 Gurkha Infantry Brigade, and part of the Squadron accompanied Robin Wilson and A Company to Malaysia for jungle training in 1971. Hearts and minds operations were also very prominent and ranged from building a school playground in Sha Tau Kok to giving blood (in 1971 alone the Regiment gave 300 pints). The 48 Gurkha Infantry Brigade Inter-Platoon Cup was another annual highlight of this period, with each platoon in the Brigade being put through a gruelling 24-hour test of their military skills. The Regiment always did well in this, and in 1971 filled the first four places in the competition. Sport, too, played its part, and under the tutelage of Akalsing Thapa and then Bhagwansing Thapa, the Sixth won the Nepal Cup in 1969, 1972 and 1973. The swimming team covered itself in even greater glory, winning the Hong Kong Land Forces Swimming Championships no less than five times over the years 1970–75. Indeed, the range of sports was wide, and there was great emphasis on making everyone participate, whatever their standard.

Apart from the departure of those who had taken the redundancy terms, December, 1970 saw Ralph Reynolds hand over command to Jim Kelly from the 7th Gurkhas. Ralph had had no easy task in handling amalgamation and the rundown, and it is a measure of his personality and skill that all went so smoothly.

In early 1971 the Regiment was warned that it would be moving to Brunei in the Autumn, although not until June was this finally confirmed, after the signing of the Anglo-Brunei Treaty. The move was due to take place in October, just before this the typhoon season kept everyone busy. Typhoon Rose, which struck Hong Kong in mid-August, caused enormous damage and many deaths, and relief operations were extensive. Typhoons even interfered with Dashera, and meant much erecting and dismantling of tents in order to prevent them from being blown away. Nevertheless, with the Advance Party flying in and the Main Body going by sea, the Regiment was firm in Seria, the old home of the 2/6th during the latter part of Confrontation, by the end of October.

Since 1966 there had been many changes in Brunei. With the closing down of the base at Singapore the Brunei battalion was now entirely dependent on Hong Kong administratively and logistically, and everything had to be flown in or go by sea. For this reason the Battalion was under command of 51 Infantry Brigade in Hong Kong. Initially, until the Brunei airport was improved, stores coming in by air had to land at Labuan, a Malaysian island off the north-west coast of Brunei, and it meant a four-hour sea crossing and three-hour drive by road before they arrived at Seria. Nevertheless, there were many compensations. The virtual independence of the Battalion, being geographically so far from its "masters", the kindness and generosity of the Sultan himself, and the chance to exercise in exotic places were but some.

The first task was to make the Regiment *au fait* with the jungle once more. Many, in spite of company training exercises in Malaysia mounted from Hong Kong, were novices, and it was very much a question of dusting off the pages of CATOM once more and getting back to basics with instruction on basha building, patrol formations, siting of sleeping places and weapons and all the old skills needed for jungle operations. Still, in their first proper jungle exercise against the Royal Brunei Malay Regiment (RBMR) and SAS in March, 1972, the Regiment acquitted itself well. An additional task was the running of six-week courses in jungle warfare for young officers from the Brigade of Gurkhas.

Brunei, being far-flung and exotic, attracted a large number of visitors, with no less than 106 during the Regiment's first year there. The highlight, however, was the visit by Her Majesty The Queen, together with HRH Prince Philip, HRH Princess Anne and Earl Mountbatten, in February, 1972. Although it was a State Visit, the Regiment was heavily involved, especially in the security of the Royal Party. There were, too, opportunities to meet Her Majesty and she seemed especially pleased to see her Pipe Banners again. For Henry Oakden and the Gurkha Major, Kholal Ale, the highlight was to have the privilege of forming part of the escort of the Royal Chariot, "an immense structure of gilded teak" which "was pushed and pulled by soldiers of the RBMR in national dress" and in which Her Majesty and His Highness The Sultan rode from the Customs House steps to the State Meeting Hall.

1973 passed in much the same way as the previous year, although the dry season was distinctly wet, a complete reversal from 1972 when the wet season was very dry. In June, Colin Scott took over from Jim Kelly, but not before the latter had to demonstrate his swimming prowess in public. Commander 51 Brigade, Brigadier P. F. A. Sibbald OBE, was visiting and watching D Company do a river crossing. Announcing that he wanted to see what was going on on the far bank, he immediately dived in and set off. Jim Kelly had no choice but to follow his example, and the two raced across the river to the cheers of the men – history does not record who won. Then came the news that the Battalion was to move back to Hong Kong in January, 1974, which was welcomed by those who yearned for the bright lights, but others, who had appreciated the peace and quiet of Brunei, were not quite so sure. Indeed, it had been a happy and rewarding tour.

The main disappointment about being back in Hong Kong was that the Regiment would not be in Gallipoli Barracks, but Cassino Lines, some seven miles to the south-west at the foot of Snowdon Hill. It was very much more spread out than the former, but was lacking a swimming pool and athletics ground, both of which had largely contributed to making life in Gallipoli so pleasant. There was also a drought, which meant strict water rationing, and the "three day week" in Britain resulted in a shortage of spares and fuel, which restricted training. However, there was one memorable exercise, SIXTH FOOT, a 48 Brigade conventional warfare scheme. Akalsing's D Company had to take up a defensive position on the eastern shoulder of Tai Mo Shan, at 2,000 ft the highest feature in the Colony. No sooner had trenches been dug and the position wired in than a typhoon signal was hoisted to mark the arrival of Typhoon Dinah. During the night D Company was subjected to its full blast, with trenches collapsing and the command post flooded. At dawn General Sir Edwin Bramall, Commander British Forces Hong Kong, and the Brigade Commander, Brigadier John Whitehead, visited and suggested that the exercise be called off. The Commanding Officer was, however, undeterred, and D Company dug a new defensive position and carried on. Another, but more pleasurable, period of training, was when Miles Hunt-Davis took a company group based on C Company to Fiji for six weeks.

1974 was also marked by Brigadier David Powell-Jones, who had commanded the 2nd Battalion so successfully during the Malayan Emergency, being appointed Colonel of the Regiment. He had retired early from the Army in order to go into industry, but his business trips had enabled him to visit and keep very much in touch with the Regiment, and the fact that he had one son, Evan, serving and the other, Philip, having recently left the Regiment for civilian life, meant that he had "his finger on the pulse" from the start. As for the Brigade of Gurkhas as a whole, a welcome occasion was the opening of the Museum by Field Marshal Lord Harding at Church Crookham in June, and the Regiment contributed various items to it.

The most significant event of 1975 was the visit of Her Majesty The Queen to Hong Kong in May. Although the tight timetable precluded a visit to the Regiment itself, the 6th Gurkhas had the privilege of providing a Guard of Honour, commanded by Major Paul Pettigrew, on her arrival, and the Commanding Officer was presented to her. Border duties continued to take up a large part of the time, as well as guards and training, but there were other activities which gave added interest. Jim Pearce and Paddy Lewis took a detachment to Korea on temporary duty with the Korean Honour Guard, something which became a regular feature of life in Hong Kong. In spite of the cold, it was an interesting experience. Korean food was found infinitely preferable to American rations, and offers of US $30 for a kukri were hard to resist. Gerald Davies also took A Company and the recruits to Brunei. In May, at the same time as the Royal Visit, there was the harrowing experience of a shipload of destitute Vietnamese refugees putting into Hong Kong. The Regiment set up a tented camp for them at Shek Kong, and the inherent generosity of the Gurkha was amply demonstrated in the mass of donations of clothing, food and toys, which were sorted and distributed by the British wives in the Regiment. A slightly unusual experience was that of Henry Hayward-Surry, who, with three pipers and two drummers, attended a mini pipe band competition in Jakarta, organized by the St Andrew's Society of Java. Although the Regiment was unsuccessful in the group events, Sgt Krishnabahadur Gurung won a gold medal as an individual.

In October, 1975, Colin Scott handed over command to Mike Wardroper of the 10th Gurkhas, having spent no less than nineteen of his twenty-four years of commissioned service at regimental duty. A similar record could also be claimed by Desmond Houston, who retired earlier in the year after thirty-six years' continuous service with the Brigade of Gurkhas. Even then he did not lose his active connection, taking over a post in the Brigade Record Office in Hong Kong. The autumn also saw the usual round of formation exercises and for the second year running the Regiment found itself on border duties over Christmas. These periods on the border had, especially because of the gradual reduction of the Hong Kong garrison, become rather more drawn out and a battalion could now expect two periods of two months each during the year. Illegal immigrants were still crossing in a steady trickle and, rather than rely entirely on static observation posts, much use was made of four-man patrols to swamp the area, especially at night.

In terms of the quality of life, 1976 saw the building and opening of a swimming pool within the Lines. Shooting was also in the news, with the Regiment qualifying to go to Bisley, where the team finished ninth in the Major Units Championship. Another noteworthy achievement in which the Regiment took part was the ascent of the 23,583 ft Annapurna South Peak. This was carried out by a twelve-man expedition from British Forces Hong Kong, which included no less than five from the Regiment, two of whom, Duncan Briggs and Cpl Rinchen Wangdi Lepcha, were among the three who reached the summit. However, early in the year, the Regiment was told that it was bound for Church Crookham in January, 1977. The prospect of UK allowances, with the chance to save more money, was welcomed but, unlike the 1/6th's tour in Tidworth, this time, because of the lack of accommodation and extra expense, the Gurkha families were not able to accompany the Regiment and they returned to Nepal.

After a farewell parade in Hong Kong, on which "Bubble and Squeak" the two old mountain guns appeared, with two riflemen dressed in the old 42nd Gurkha Light Infantry uniform, the Regiment began its move in February. Gordon Corrigan took B Company to Sennybridge in order to prepare for its role as the Demonstration Company at RMA Sandhurst, while the Advance Party began a lengthy take-over of the barracks at Church Crookham. The Main Body flew from Hong Kong in April, and one of the first social occasions was when a representative body went down to Bovington at the invitation of the 14th/20th King's Hussars who were then the Royal Armoured Corps Centre Regiment. A month later there was a memorable Regimental Weekend, with many famous names from the Regiment's past attending, including Aileen Viscountess Slim, Field Marshal The Lord and Lady Harding, General Sir Walter and Lady Walker, Lt Gen Sir Reginald Savory, who, as Adjutant General India, had been responsible for supervising the transfer of the Brigade of Gurkhas to the British Army in 1948, Major General "Punch" Cowan, the wartime commander of 17th Indian Division, the "Black Cats", and his wife and Major General and Mrs Jim Robertson.

1977 was the year of the Royal Silver Jubilee and it was fitting that the Regiment should be in England to celebrate it. Their role was to line the route at Temple Bar, where Her Majesty's coach halted for her to receive the Sword of the City of London. The Pipes and Drums were also in heavy demand for various Jubilee celebrations, and a small party, including the Gurkha Major, was privileged to attend the Royal Naval Review at Spithead on board HMS *Gurkha*. Bob Richardson-Aitken and Lt (QGO) Bhadre Thapa took another party in her later in the year when she sailed from Liverpool to Rosyth for a refit. After the initial seasickness wore off, everyone thoroughly enjoyed themselves, especially firing the 0.5 Oerlikon AA gun at balloon targets.

One of the major commitments of the Church Crookham battalion was that of running Bisley, and this took up much of the summer. Having won the South-East District Major Units Championship, the 6th's place at Bisley was assured. In the main competition the Regiment finished third, behind the 2/2nd and 10th Gurkhas, and in the Individual Rifle Championship five men were in the Army Hundred, including Rfn Dharmendra Gurung who finished eighth. As for the running of Bisley, this involved not just the Regular Army Meeting, but TAVR, National Rifle Association and Cadets as well; the commitment was almost 400 men for two months. Under the command of Evan Powell-Jones, they received high praise for their efficiency, enthusiasm and cheerfulness, and many said that it was the best-run Bisley ever. Then followed the honour and pleasure of taking over the guards at Buckingham Palace, St James's and the Tower of London from 1st Bn Scots Guards. This necessitated the provision of properly tailored No 1 Dress for all those involved, and much of the credit for their very smart turnout was due to Major Reg Creamer, the Quartermaster. There were, of course, moments of light relief amid the solemnity of it all, like the time that Capt (QGO) Dalbahadur Gurung, commanding the Buckingham Palace Guard, was somewhat mystified to have a ten-page grievance thrust into his hand, or Manikumar Rai at the Tower, who, while overtly disbelieving all tales of ghosts and spectres there, nevertheless insisted on keeping his light on all night.

At the end of the summer, in order to have a change of scene, a series of company exercises took place in Cyprus. Highlight of this was the Cyprus Walkabout, an annual competition involving a sixty-mile hike through the Troodos Mountains in teams of three, and the Battalion's A Team did well to finish second. Everyone was back in Church Crookham for Dashera, although the celebrations were not the same without the families. Another event in October was when the Commanding Officer, Mike Wardroper, Mike Whithead, the Second-in-Command, and Gurkha Major Toyebahadur Chand were invited to the Annual Dinner of the Gallipoli Association in London, together with the Colonel of the Regiment. At it they heard General Savory pay particular tribute to the 1st Battalion's part in the action at Sari Bair in 1915.

Another period of Public Duties was looming up in December, but in early November, just when training for it was starting, the Battalion received a warning order for an entirely different task. At this time the firemen were about to be called out on official strike over pay, and it was decided that the Armed Services would have to step in to fill the breach. The order called for three rifle companies to be mobilized and, within four days, they had been rushed down to the Royal Navy's firefighting school at HMS *Excellent*. Here they were introduced to the famous "Green Goddess", a somewhat antique (by Fire Service standards) fire engine, and learned to operate the hoses, extinguishers and breathing apparatus. Then they returned to Church Crookham to receive their own Green Goddesses and, with the firemen intending to go on strike from 0800 hours the next morning, they deployed on the evening of Sunday 13 November. The Battalion was given responsibility for Berkshire (Support Company) West Sussex (D Company) and Surrey (C Company). In all twenty-two Green Goddesses were manned, each by a section, which supplied two crews of a commander (officer or senior NCO) and five men for each shift, which was twenty-four hours. It was not possible to use the regular fire stations because of the strike, and the Green Goddesses were based at TA drill halls, Regular Army barracks and even hospitals. Crucial to success was the basing of a fire officer and policeman at each location, the former to give advice and continuation training and the latter to clear the way when the fire engines were called out. Communications were also important and relied on the Fire Service nets, police and the normal company radios, with each company headquarters working directly to the County Emergency Headquarters.

Support Company had only a week in Berkshire, based at Reading, before they were called back to Church Crookham to provide part of a nationwide emergency stand-by force. They had no major fires, and their main problem was getting the Green Goddesses fitted with blue lamps as a warning to motorists travelling at high speeds in conditions of poor visibility on the M4 Motorway – the bronze green fire engines proved to be too well camouflaged and there were some alarming near misses on their first emergency call-out. D Company, which was an amalgam of A and D, together with the Pipes and Drums, had three weeks in West Sussex, with their base at the Royal Military Police Training Centre at Chichester, and dealt with some seventy fires ranging from electrical fires in domestic appliances to a factory fire. However, Public Duties in London still had to be carried out and were due to begin on 4 December. So they returned to Church Crookham a week before this in order to prepare.

This left Major Gopalbahadur Gurung and C Company, and they remained in Surrey for nine weeks until the end of the strike. Bearing in mind that Surrey normally has forty well-equipped fire engines and some 700 firemen and Gopal had only twelve Green Goddesses and 150 men, C Company was very stretched. Nevertheless, they answered a total of 643 calls and very soon gained a high reputation for their dedication and cheerfulness. This was well demonstrated over Christmas, when each base was literally swamped by gifts of food and drink. One notable visitor to the Redhill base was the actor Oliver Reed, who brought two bottles of whisky with him. Off-duty crews did have the opportunity of getting back to Church Crookham, and sampling the local social life. One Gurkha recalls: "We were called out when the disco was in full swing and we were showing off our Gurkha/English twists to the WRAC girls. Trust our luck!" Even on duty there were lighter moments, as when Bhadre Thapa, fighting a fire at the Weybridge Club, became so incensed at the failure of his hose to break a window that he threw it down, drenching everyone in range, and with an unintelligible oath heaved a brick through the window. There was also the occasion, which was mixed with pathos, when a crew was called out to rescue an old lady's kitten stuck up a tree. Rescue the kitten they did, but only to run it over when they were backing out of the drive. At the end of the strike, in order to mark their appreciation, the Surrey County Council gave a formal dinner for C Company and presented them with an inscribed fire bell. Later recognition of their work came in the form of an MBE for Gopalbahadur Gurung. Finally, after a mounted parade at Blackdown to bid farewell to their Green Goddesses, C Company were allowed to stand down, but were given permission to keep their firemen's helmets. As Major Mike Whitehead wrote; "You never know what you might see in the hills of West Nepal in years to come".

By now it was 1978 and the most exciting event in the diary was the visit by Her Majesty The Queen to Church Crookham at the beginning of May. Preceding this, Major Patrick Gouldsbury, the unit Press Officer, had been busy, and managed to persuade BBC South to produce a 30-minute programme on the Regiment, which was screened a week before the Royal Visit. He also, among other "scoops" that year, got BBC South to do a piece on the forty-six sets of brothers serving in the Regiment at the time. The day itself, 5

May, alas was wet, but this did not detract from the joy of the occasion. Her Majesty saw the Battalion both at work and play, was entertained to *bhat*, at her own request, in the Officers' Mess and among others met Tulbahadur Pun VC, who was in England for a VC/GC reunion. Finally, after being entertained to some traditional dancing in the Dashera Ghar by the nautch party, she departed in the Royal Range Rover, led by the Pipes and Drums playing the Regimental March and accompanied by cheering soldiers throwing flowers into the vehicle. As the report in *The Kukri* said: "Thus ended a most exciting day which even the most miserable weather could not spoil, and one which those of us who were present will never forget."

The next major event was Bisley, at which the Battalion was determined to improve on the results of the previous year. However, before this and just after Her Majesty's visit, a composite company under Major Akalsing Thapa spent an interesting three weeks in Norway as guests of the Royal Norwegian Army. Although the riflemen did not take too kindly to the Norwegian diet of tinned meat, fish, sausage and bread, *bhat* became very popular with their hosts. Later came the South-East District Skill at Arms Meeting, in which the Battalion collected almost every trophy that was available, but at Bisley, which the Battalion was again running, it could manage no better than third place for the second year. There was, however, consolation for the Shooting Team in that they were invited to go immediately to Canada to represent the British Army in the Canadian equivalent, which was held at the Connaught Ranges near Ottawa. With Major Brian O'Bree as Adjutant, they won a number of prizes, Rifleman Rajendra Gurung winning the Sir Arthur Currie Gold Medal.

However, while Bisley was taking place, the Battalion was suddenly told that it was to deploy to Belize for a seven-month tour in three weeks time. The news came just as Mike Wardroper was handing over command to Christopher Bullock, who had come from the 2/2nd. For the latter it was a perplexing start to his time in command, partly because the Battalion was new to him, but also because he was immediately given twenty-four hours' notice to fly to Belize on a recce. Thus, leaving Mike Whitehead to extricate the Battalion from Bisley and hand over its duties to the Royal Irish Rangers, who had been originally scheduled for Belize, Christopher Bullock set off. He was accompanied by the Operations Officer, Ian Bushell, and the Education Officer, Keith Boulter, who was to act as Intelligence Officer.

When British Honduras gained her independence in 1975 and

became Belize, she was immediately faced with the problem of an external threat from her western neighbour Guatemala. The latter had had a claim on the territory of Belize since the mid-19th century, and Guatemalan maps showed it as a province of their own country. After Independence, the Guatemalans concentrated their attentions on the southern part of the country, Toledo District, and made threatening military moves on the Border. Belize had nothing in the way of defences, except for her weak Defence Force, and it was decided that British troops should be deployed to guard against Guatemalan incursions. This Force was built around two all-arms groups each based on an infantry battalion. They were called Battle Groups North and South and were supported by some Harriers and a Royal Navy frigate. The task of the 6th Gurkhas would be to take over from the Royal Highland Fusiliers in Battle Group South and be responsible for some 1,800 square miles of territory. They were to be the first Gurkha troops to be stationed there.

After the return of the Recce Party, there was a mass of work to be done, but very little time in which to do it. As Christopher Bullock recalled:

"Since Belize battalions expected six months' notice before a Belize tour I felt that three weeks was cutting it fairly fine! However, although I didn't know the 6th, I knew Gurkhas and I was sure that they could cope with it given a firm steer. This I gave at a briefing I held on the day after I got back. Since I was limited to 350 men I had decided to take A Company, commanded by Major Gerald Davies who had just come back from the Oman where he had seen action, and B Company, commanded by the experienced Major Gordon Corrigan. . . . After the two rifle companies were totalled there was only headroom left for a small support company of two 81 mm mortar sections and two anti-tank gun detachments. Battalion Tactical Headquarters, the QM and a small staff, MT and Signals were all kept to a minimum."

Even so, with the Demonstration Company at Sandhurst still to run, together with maintaining Church Crookham, the Battalion was very stretched and the Commanding Officer was forced to ask HQ Brigade of Gurkhas for the thirty men at the Battle School at Brecon to be relieved. HQ Brigade of Gurkhas could not have been more sympathetic, and very quickly a relief party from the 10th Gurkha Rifles, then in Brunei, flew over, and a troop of Gurkha Engineers was also offered and gratefully accepted.

The Battalion flew to Belize over the period 4–8 July, going via

Gander, Newfoundland. Contrary to the normal practice whereby battalions had a month to acclimatize themselves to the jungle before becoming operational, the Commanding Officer, with many of the Battalion already having jungle experience, volunteered to dispense with this. Consequently A Company deployed immediately, although they found the jungle was, by Far Eastern standards, particularly "dirty", with a multitude of snakes, scorpions and tarantula spiders, apart from the jagged limestone "Karst", which was very difficult to traverse. Battalion HQ and B Company were based at Rideau Camp, some 100 miles south of Belize City, on the coast. Here B Company's task was to counter any sea or airborne incursion from the east, while A Company with its base at Salamanca Camp, twenty miles north-west of Rideau in the jungle, were to cover against any attack through the jungle from the west. In order to assist the Battalion, they were given under command a troop of three 105 mm light guns from 7 (Sphinx) Commando Battery RA and a Blowpipe surface-to-air missile detachment, and there was also a Royal Engineers troop on hand, besides the Gurkha Engineer troop. The latter were particularly useful, since the camps were in a dilapidated state, apart from the fact that, as the Commanding Officer had noted on his recce:

"I was struck by how much the British Army had forgotten since Borneo a bare fifteen years before. None of the camps were defended, they were all poorly maintained and the tents and buildings were unsuitable and unnecessarily uncomfortable. For instance the huts were steel Tynehams and, devoid of shade, became stiflingly hot."

Both Rideau and Salamanca, along with Cattle Landing, just by Rideau Camp, where Support Company was based, were quickly brought up to scratch:

"Soon we had a properly dug-in defensive system which started with observation and machine-gun posts on the surrounding hills and ended with wire and bunkers round the camp. Inevitably in the wake of this martial building came the little gardens, briefing centres, shady attap huts and games patches so beloved of the Gurkha."

Once the initial period of jungle training had been completed and the camps made more secure and comfortable, patrolling with live ammunition in the jungle began. This, until the Battalion's arrival, had been very much the preserve of the SAS, but they quickly acknowledged the Battalion's professionalism and returned to the United Kingdom. Life was not made easy by the fact that it was the wet season, and in the south of the country this can amount to as much as 170 inches per annum. The main effect, however, was to make land communications difficult, and much of the Battalion's resupply was dependent on Puma helicopters and lighters. Although the patrolling was long and arduous every Gurkha managed to get to the beautiful island of Placentia off the Belize coast, with its golden beaches and friendly girls, while many of the officers made trips to the Caribbean and USA.

During the last three months of the tour the Guatemalans became restive and indulged in one of their periodical bouts of sabre rattling. Patrolling of the border was therefore stepped up and the number of observation posts increased. Inadvertent border crossings were naturally enough frowned upon, but there was one incident over Christmas when an RAF Puma carrying Sgt Krishnabahadur Chhetri and ten men of A Company deposited them half a mile on the wrong side. Nothing daunted, the party charged through a band of Guatemalan military engineers building a road up to the border and made their way quickly back to friendly territory. Besides operational patrolling, the Battalion carried out a number of exercises, including one before the Commander Belize Garrison, Colonel Angus Robertson, to demonstrate a new contingency plan should the Guatemalans attempt incursions through the jungle. This was very successful, so much so that it is in use to this day. Indeed, the tour as a whole had more than demonstrated the Battalion's calibre and there is no doubt that as a result of their performance it has been policy ever since to station the Church Crookham battalion in Battle Group South every other six months. As for the 6th Gurkha view of it all:

> "From Bisley to Belize was certainly a change. Although as yet no violent revolutions and very few exotic ladies have come our way – daily working with Pumas, simulated airstrikes by Harriers and operational jungle patrolling most certainly have; and for these fantasies coming true we are grateful. It has been nice to have a real job to do once again."

And so the Battalion returned to the English winter at Church Crookham, where there had been some changes in personalities. First and foremost, Brigadier David Powell-Jones had come to the end of his five years' tenure of the Colonelcy of the Regiment, and it was sad to see him go. However, in his place came Brigadier Sir Noel Short MBE MC. Brunny Short's name, like that of his predecessor, has appeared many times in these pages and he had, after retiring in 1964, joined the Civil Service, rising to become Secretary to the Speaker of the House of Commons. It was good to know that he would be with the Regiment again in an official capacity. However, one face which would be sorely missed was that of Henry Hayward-Surry. He was the last to serve with the Regiment who had fought during the War, when he had been a Chindit with the 3/4th, before transferring to the British service when Independence loomed after the War. He had joined the 1/6th in 1950, seeing much active service both in Malaya and Borneo. His ready and sharp wit and deep knowledge of the Gurkhas and their homeland had made him an indispensable member of the Sixth "community". The Pipes and Drums owed him a special debt for his care of and interest in them.

Arriving back from Belize in February, with the knowledge that they were to return to Hong Kong in May, gave the Battalion little time to pause for breath. Once again the move meant a change of role, and besides packing up and preparing to hand over the Battalion's responsibilities, retraining in internal security drills and tactics was necessary. However, on arrival, a different task confronted the Regiment. Red China was going through one of its periodic upheavals and the result was a flood of people trying to slip across the border into Hong Kong. Indeed, the Advance Party found itself immediately committed to the border, and the Commanding Officer himself, whilst visiting his old Battalion witnessed "a group of about a hundred refugees burst over the border in front of us. Myself, orderly and signaller managed to catch eight and 2/2 GR patrols who were spread very thinly a few more, but the majority disappeared into the foothills around Fanling." Thus, the main party hardly had time to settle into its new home, the picturesque little camp at Burma Lines, Fanling, when it was deployed on border duties. Any hope of allowing the men to be reunited with their families, especially those away from Nepal for the first time, vanished, and it was left to the Gurkha Major, Dalbahadur Gurung, to settle them in.

The Battalion's area of responsibility was the extreme west of the New Territories, covering the coastline from Castle Peak to Deep Bay. Gopalbahadur Gurung with C Company held the key area of the Maipo marshes, and found themselves catching up to 200 Illegal Immigrants (IIs) every twenty-four hours. B Company (Ian Bushell) was in the mid-Castle Peak area around Nim Wan, and D

Company (Mark Harman) was in Lau Fau Shan. Peter Degraaf, who was on a two-year exchange from the Royal Australian Regiment, with A Company was nominally in reserve, but more often than not deployed as well. As had become the practice during the Battalion's last tour in Hong Kong, the basic unit was the four-man "brick", consisting of a Lance Corporal, radio operator and two riflemen. Between them they had a rifle, a Federal Riot Gun, which fired baton rounds, a stick each and their kukris. By day comparatively few IIs tried to cross, which enabled the men to get some sleep, with just observation posts and a couple of quick reaction teams standing by with a helicopter on call. However, once darkness fell it was, in the Commanding Officer's words,

> "a different scene as the majority of our manpower deployed to coves, inlets and vantage points to help counter the flood of illegal immigrants who tried to break in. They swam, came crammed in little boats, pushed sledges and scooters over the mud and floated in as hopeless corpses killed by cold or sharks. Women looked much the same as men, bedraggled, wet mud-stained creatures in blue denims with bare feet ripped and bleeding on the rocks and shells. Some came carrying children, others heaved aged relations all bent on sampling the good life and all determined to try again even if caught."

Most IIs accepted their fate – handing over to the police and repatriation over the Border – philosophically. Occasionally, though, there was a violent customer, armed, as often as not, with a home-made knife, or he might be a martial arts practitioner, and this resulted in the odd injury. For the Gurkha, who is very much a family man, it was often a harrowing task, but, like everything else required of him, he accepted that it had to be done and got on with it to the best of his ability.

During the first month alone, from the end of May to the end of June, 1979, the Battalion caught no less than 3,600 IIs, and so it went on month after month until the Battalion was finally, in February, 1980, able to have a break. During the early months all four battalions (three Gurkha, one British) in the Colony were deployed full time against the IIs, but the arrival of Brigadier Ian Christie to take command of the Gurkha Field Force* in December, 1979,

---

* 48 Gurkha Inf Bde became Gurkha Field Force as a result of the British Army 1974 reorganization, which abolished the designation of Brigade, but retained this title when "brigades" were readopted in 1980 so as to avoid confusion with the Brigade of Gurkhas.

brought about a gradual change in plan. Clearly, with no time for anything else but border security, training was suffering, and since there appeared to be no immediate end to the problem in sight, fresh measures had to be taken. A continuous fence was constructed across the land border and a fresh battalion flown out from England on a four-month emergency tour. These steps enabled two of the resident battalions to be stood down.

B Company made the most of this break by dashing off to New Zealand for six weeks, where they exercised with the 2/1 Royal New Zealand Infantry Regiment. The remainder took part in the annual Colony internal security exercise SPRING TROT. But, all too soon, they were back on the Border again, where the Battalion remained for a further three months. It was noted that the IIs still managed to get through the new fence, although it took them considerably longer than with the old primary fence. However, in October, the Governor of Hong Kong declared an end to the traditional "touch base" policy, whereby an illegal immigrant was issued with identity and right of residence once he had reached the urban areas and found accommodation. Within a matter of days the daily number of illegal border crossers dropped from four hundred to four. The crisis was finally over, but border duties continued to occupy the Battalion for the remainder of its tour in Hong Kong.

January, 1981, saw Christopher Bullock hand over command to Paul Pettigrew. His tenure of command had been the busiest the Battalion had experienced since amalgamation, and it was a measure of both Christopher Bullock's leadership and the Battalion's response that he was made an OBE. 1981 and 1982 did provide more opportunity for sport and recreation. Nevertheless, Border duties continued, and the Battalion's final total bag, when they were stood down on 1 September, 1982, was no less 15,141 illegal immigrants apprehended. However, it was the Battalion's sporting achievements which really came to the fore, and in 1982 they were the Hong Kong Khud Race,* Swimming, Athletics and Cross Country League champions. The Shooting Team also covered itself in glory, finally winning the Major Unit Championship at the

---

* The Khud Race is a relic of the campaigning by the old Indian Army on the North-West Frontier of India. The key was domination of the high ground, which meant getting picquets up onto it quickly. Conversely, when the column moved on, it was equally essential for the picquets to be withdrawn quickly, or otherwise they would get cut off by the enemy tribesmen. Much training was therefore carried out in racing up and down steep hills, and this became a sport as well. Gurkha and British units take part in an annual Khud Race in Hong Kong.

Army Rifle Association Meeting at Bisley in 1982 with a record score, Cpl Dharmendra Gurung winning the Queen's Medal as the British Army's champion shot. They repeated this success at the National Rifle Association Meeting and went on to win further trophies at the Canadian Bisley.

Prior to departure from Hong Kong for Brunei in November, 1982, it was decided to hold a Regimental Reunion over Dashera, and it proved to be a memorable one for all concerned. From England came Henry Hayward-Surry, Colin Scott and Roy Pavey (a wartime 6th Gurkha), from the USA John Conlin, from Australia Neil Anderson, Ian Peters and Mike Channing, and resident in the Colony itself were Desmond Houston, Lawrence Pottinger, Jack Keen and Gordon Corrigan. However, from Nepal came no less than twenty-four former members of the Regiment with almost a thousand years' service between them. There were Tulbahadur Pun VC, Tekbahadur Subba, Hariprasad Dewan, Harkasing Rai, Kholal Ale, Tambasing Gurung, and many more whose names will always crop up in conversation when two or more 6th Gurkhas gather together. The fortnight seemed one endless long party, with "Do you remember?" on everyone's lips, but there were highlights. The celebration of Kalratri in Gurkha Major Jaibahadur Gurung's magnificent Dashera Ghar, where that veteran *maruni* Capt (QGO) Birbahadur Thapa showed that he could still teach the young dancers a thing or two. Then there was the steadiness of the recruits at the Passing Out Parade at the Training Depot, and the badging ceremony afterwards at which each recruit was presented with his Regimental Badge by a *buroh sahib*, a symbolic handing on of all the Regiment stands for, from those who have passed before to those poised to carry its spirit and traditions into the future. Finally, there was the Beating of Retreat, which marked the Battalion's official farewell to Hong Kong to begin a new chapter in the life of the Regiment in Brunei.

As the Evening Hymn was played, the Last Post sounded and the Lone Piper played "Sleep Dearie Sleep", many in the audience must have reflected on the past, present and future of the 6th Queen Elizabeth's Own Gurkha Rifles. They may have thought back to the early days of the Regiment, its raising as the Cuttack Legion, its long service in Assam and on the NE and NW Frontiers of India, Gallipoli and Mesopotamia during the First World War, further Frontier campaigning between the wars, and Burma and Italy during 1939–45. Thereafter come more recent years, with the agonies of the opt, the beginning of a new life and the long struggle in Malaya, followed by Confrontation in Borneo, that first tour in England, the numerous sojourns in Hong Kong, Brunei, amalgamation, the second visit to England, with fire fighting and Belize, and the many other places which detachments, large and small, have visited. Throughout all hangs a constant thread, the courage, devotion and uncomplaining loyalty of the hillman from Western Nepal, who is the Regiment. Whatever the challenges of the future, they will be met with the same steadfastness that the Gurkha soldiers of the Regiment have shown in the past, and the history of that past will act as inspiration for the future.

Annexure 'A'

AN ACTION AGAINST A COMMUNIST TERRORIST CAMP IN THE BATU GAJAH AREA OF PERAK IN OCTOBER, 1953

Background

In October, 1953, the battalion responsible for the Batu Gajah district of Perak was temporarily deployed elsewhere and for a short time this district was added to the area for which 1/6 GR were responsible. Late one evening a Communist Terrorist (CT) surrendered to the Police with vital information concerning an important meeting between the State Committee Member and five District Committee Members of the CT organization. The meeting was to be held in the jungle not far from Batu Gajah town, when the District Committee Members would hand over their half-year collection of funds amounting to some half a million dollars.

At the time that the CT surrendered, the whole of 1/6 GR was already deployed except for the reserve company (A Coy) and one platoon of D Coy. No British Officer was immediately available, as OC A Coy was away on a course, but OC D Coy (Maj J. A. Lys) was just completing local leave following his honeymoon in the Cameron Highlands, so he was summoned to react to the information with two platoons of A Coy and one platoon of his own D Coy.

Interrogation of the surrendered terrorist (SEP) had been thorough:

a. The total number of CTs was 7, i.e., the 6 VIPs plus one sentry.
b. The SEP was familiar with the various routes to the camp and was willing and able to lead the Company to the camp by night by a route which would avoid the sentry.
c. The 6 VIPs slept separately within a comparatively clear area of jungle of about 15 yards radius.

d. The Meeting would split up early the next day. To achieve surprise and success, the attack had to be made at first light with vehicular movement and the approach march after dark.

Preparation
Everything possible was done to eliminate noise prior to leaving the barracks, e.g.
a. All spare magazines were wrapped in cloth.
b. Bayonets, metal torches and match boxes were discarded.
c. Men with coughs and colds were left behind.
d. All magazines were fully loaded to avoid rattling. (All LMG magazines were loaded with 100% tracer to aid the aim of the firer at night.

Approach march
The Company reached the debussing point at approximately 2200 hours. It was estimated that at normal walking rate the camp could be reached in about two hours in daylight. It was, therefore, thought that a silent approach at night would take about 6 hours. In fact, it took the Company 7½ hours, of which at least 4 hours was spent walking through waist-deep swamp.

A temporary base was established about 1000 yards away from the CT camp with a platoon sergeant in charge. Large packs and equipment not wanted for the assault were left at the base with the radio operators plus some further riflemen identified as "coughers".

Organization for the assault
The company was organized into an Assault Group and a Cut-Off Group. Further interrogation of the SEP revealed that the Assault Group, in order to avoid the sentry, would have to approach the perimeter of the CT camp over a vegetable garden 30 yards wide by about 50 yards long. Maj Lys was, therefore, forced to restrict the assault group to one Platoon Commander, flanked by 2 LMGs, himself flanked by 4 LMGs plus his batman. The remainder were allocated to the Cut-Off Group under command of Capt (QGO) Manbahadur Gurung (III) of 'A' Company.

The plan
The plan was to proceed from the Base Camp to the edge of the vegetable garden in single file with the 'O' group in front followed by the Assault Group and the Cut-Off Group. From this point it was hoped the SEP could point out exactly where the CT camp was, to

facilitate the deployment of the Cut-Off Group. Both groups were to be in position by 0430 hours and the commander allowed four hours to complete this phase.

The assault

The edge of the vegetable garden was reached in complete silence. Maj Lys ordered the SEP to guide the Cut-Off Group around the sentry to cover all escape routes. Half an hour later the Assault Group crawled along the furrows between the vegetable mounds to the very perimeter of the CT camp. The timing was almost exactly right and complete surprise was achieved. One of the CT's huts was just 20 yards from the Assault Group. The signal to "open fire" was to be given by Major Lys by touching the backs of the Bren Gunners on his left and right. Just after first light, when four of the six CT huts could be seen, Major Lys observed a CT under a mosquito net sit up and yawn and stretch his arms, and he immediately gave the signal to open fire.

Firing continued for about 5-6 minutes. All six VIPs were killed and identified by the sentry escaped. Regrettably no money was found although the Company searched the area until 1100 hrs.

Conclusion

The operation had been a complete success. Swift reaction to important information coupled with a high standard of training and jungle-craft had resulted in the elimination of the CT hierarchy in the Batu Gajah area.

For his leadership and skill during this operation, Maj Lys was awarded the Military Cross.

Hong Kong, 1979 – Another batch of illegal immigrants for return across the border.

The Steadfast Gurkha.

Annexure 'B'

1968, HONG KONG. AN INCIDENT AT MAN KAM TO

D Company of the 1/6th under Richard Lowe were involved in a tricky incident at Man Kam To. As part of the Hong Kong Government's plan to close the border to all traffic, pedestrian or vehicular, a barricade was erected on the Hong Kong side of the bridge crossing at Man Kam To. This barricade consisted of a metal sheet about half an inch thick and twelve feet high completely sealing the bridge, a sandbag bunker housing a section of men on the Hong Kong side, and large quantities of dannert wire on the bridge itself. The orders to the section commander in the bunker were to prevent the Chinese from entering Hong Kong across the bridge and from removing the barricade. On a quiet peaceful Sunday afternoon, three of D Company's OPs simultaneously reported three separate groups of civilians and militia marching with banners towards the bridge. In total approximately six hundred Chinese assembled on the Chinese side of the bridge and proceeded to demonstrate, demanding that the bridge be opened so that trade from China to Hong Kong could continue. Gradually the situation escalated to the point where the Chinese began to dismantle the barricade. A 48 Brigade helicopter reported that there were women and children in the crowd and orders were passed from 48 Brigade not to throw CS gas, but to hold the position. The position for the section in the bunker was beginning to become untenable as hand-to-hand fighting broke out. Lowe ordered CS gas, but the Chinese donned masks and neutralized the thrown CS gas canisters in buckets of water. He therefore ordered white phosphorus grenades which successfully dispersed the demonstrators. The incident ended after about four hours, but not without A and C Companies being brought forward from Gallipoli Barracks and several retaliatory high explosive stick grenades from the Chinese Communist Army; one rifleman in C Company was slightly wounded. Corporal Narbahadur Pun of D Company, the bunker commander, displayed excellent determination and devotion to duty, throughout the incident, in holding the bridge under very trying circumstances.

Glossary of Gurkhali/Malay
Terms Used in the Text

Bans   Bamboo

Bhat   Cooked rice

Bhat pakaune   Cooking the rice

Bhoj   A small celebration, usually on a company basis, including drinking and dancing.

Buroh Sahib   An old veteran, British or Gurkha Officer

Chippy alus   Potato crisps

Dal   Lentil

Dashera   The main annual Gurkha religious festival, which takes place over ten days.

Dashera Ghar   The structure, usually tented, in which the main Dashera celebrations take place.

Gunong   Malay for "hill" or "mountain"

Jirgah   A conference of elders (derived from the North-West Frontier of what was then India)

Jhola   Haversack

Kalratri   The eighth day and climax of Dashera

Ladang   Malay for a cultivation in the jungle

Maruni   Male performer at a *nautch*, but in female dress.

Milap   Communication

Nautch   Dance of a more formal kind than the solo exhibitions at a *bhoj*.

Pun   One of the Western tribes of Nepal freely recruited by the Regiment

Sunao'd   Given out or explained.

APPENDIX ONE

Roll of Honour

1948–1982

Malaya 1948–60

21144683 LCpl	Dhawa Ghale	2/6th	17 Jul 48
21134142 Rfn	Gopilal Gurung	1/6th	26 Jul 48
21134375 Rfn	Motilal Pun	2/6th	3 Aug 48
398035 Lt (KGO)	Randhoj Gurung	1/6th	31 Aug 48
21144559 Rfn	Ratnabahadur Pun	1/6th	31 Aug 48
366368 Maj	R. G. Barnes	1/6th	13 Jan 49
388554 Capt (KGO)	Tulparsad Pun	1/6th	13 Jan 49
21134272 Rfn	Tarachand Thapa	1/6th	13 Jan 49
21143579 Rfn	Lalbahadur Pun	1/6th	13 Jan 49
21144062 Rfn	Ambersing Thapa	1/6th	13 Jan 49
21143855 Rfn	Balsing Rana	1/6th	13 Jan 49
21135309 Rfn	Beljang Gurung	1/6th	13 Jan 49
21144378 Rfn	Chhabilal Thapa	1/6th	13 Jan 49
21144152 Rfn	Dilbahadur Pun	1/6th	13 Jan 49
21134044 LCpl	Gaine Gurung	1/6th	13 Jan 49
21143935 Rfn	Girsing Chhetri	1/6th	13 Jan 49
21144063 Rfn	Damarsing Thapa	1/6th	11 Mar 49
21144157 Rfn	Bombahadur Gurung	1/6th	12 Apr 49
21135010 Sgt	Tekbahadur Ale	1/6th	17 Nov 49
21136296 Rfn	Patiram Thapa	2/6th	5 Jun 50
21136437 Rfn	Sukbahadur Gurung	2/6th	16 Aug 50
21144447 Rfn	Lilbahadur Thapa	1/6th	24 Aug 50
21134917 Rfn	Lachhimiparsad Gurung	1/6th	25 Dec 50
21145259 LCpl	Hirasing Gurung	1/6th	20 Feb 51
21136965 Rfn	Tikaram Thapa	1/6th	20 Feb 51
399595 Lt (KGO)	Dhanbir Thapa	2/6th	18 Mar 51
21134450 Sgt	Durge Thapa	2/6th	18 Mar 51
21136368 Rfn	Nilbahadur Khan	2/6th	29 Mar 51

388589 Capt (KGO)	Santabahadur Gurung IDSM	1/6th	10 May 51
21135106 LCpl	Pahalman Pun	1/6th	26 May 51
21136308 Rfn	Padamsing Gurung	1/6th	26 May 51
21136944 Rfn	Dhupdarja Tamang	1/6th	19 Jun 51
21134039 Rfn	Pahalman Gurung	1/6th	19 Jun 51
21135034 Sgt	Maitasing Limbu	1/6th	13 Jul 51
21143214 Rfn	Tekbahadur Rai	2/6th	12 Oct 51
384999 Maj	A. J. D. Macdonald MC	1/6th	3 Jun 52
21145599 Rfn	Indrabahadur Gurung MM	1/6th	21 Oct 52
21135203 LCpl	Ganbahadur Pun	1/6th	27 Oct 52
21134030 Rfn	Kharkabahadur Gurung	1/6th	27 Oct 52
21136350 Rfn	Amarsing Thapa	1/6th	14 Apr 53
21133750 Rfn	Purnabahadur Thapa	2/6th	4 Jun 53
21136428 Rfn	Sherbahadur Gurung	2/6th	4 Jun 53
21131866 Rfn	Lalbahadur Gurung	2/6th	1 Sep 53
21136448 Rfn	Sukdeo Gurung	2/6th	17 Apr 54
21134469 Rfn	Lalbahadur Gurung	1/6th	20 Jul 54
21136045 Rfn	Tulparsad Rana	2/6th	3 Aug 54
21136070 LCpl	Deobahadur Ale	2/6th	17 Aug 54
21134172 Sgt	Bhagtabahadur Gurung	1/6th	30 Sep 54
21136397 Rfn	Balbahadur Thapa	2/6th	28 Jun 55
413969 Capt	G. N. B. Hart	1/6th	3 Jul 55
21136089 LCpl	Satalsing Gurung	2/6th	13 Jul 55
21134555 Rfn	Purnabahadur Ale	1/6th	26 Dec 55
21138512 Rfn	Palbahadur Gurung	2/6th	12 Mar 56
21139718 Rfn	Sangaparsad Thapa	2/6th	5 Dec 56
21138607 Rfn	Madbarsing Bura	1/6th	28 Dec 56
440745 Lt (QGO)	Pemba Tshering Lama	2/6th	29 Jan 57
21134510 Rfn	Manbahadur Gurung	2/6th	8 Jun 57
21143248 Rfn	Minbahadur Mall	2/6th	8 Nov 58
21142667 Rfn	Parsad Gurung	2/6th	14 Feb 60

Borneo 1963–66

459337 Lt	F. H. Wallace	2/6th	16 Aug 63
21151891 Rfn	Lachhinbahadur Pun	1/6th	18 May 64
21150564 Rfn	Harising Gurung	1/6th	18 May 64
21143599 WO2	Chandrabahadur Thapa	1/6th	21 Jun 64
21153372 Rfn	Dattabahadur Thapa	1/6th	21 Jun 64
21153386 Rfn	Indrabahadur Gurung	1/6th	21 Jun 64
21153334 Rfn	Ombahadur Ghale	1/6th	21 Jun 64
21148319 Rfn	Dud Ghale	1/6th	21 Jun 64
21144208 Sgt	Karnabahadur Gurung	2/6th	11 Dec 64
21150613 Rfn	Lachhinbahadur Gurung	2/6th	11 Dec 64
21148328 Rfn	Jaganraj Gurung	2/6th	11 Dec 64
21149471 LCpl	Birkhaman Ghale	1/6th	24 Jun 65

21153479 Rfn	Lachhiman Gurung	2/6th	5 Aug 65
21153474 Rfn	Krishnabahadur Gurung	2/6th	5 Aug 65
21156179 Rfn	Chandraprasad Pun	2/6th	17 Jul 66
21155060 Rfn	Narbabahadur Pun	2/6th	17 Jul 66

Honours and Awards

1948–1982

Ranks and Decorations are shown as at the time of the award, and individuals are listed chronologically under each award.

KCB
Lt Gen W. C. Walker CB CBE DSO
CB
Maj Gen J. A. R. Robertson CBE DSO
Maj Gen W. C. Walker CBE DSO
Maj Gen A. G. Patterson DSO OBE MC
CBE
Brig J. A. R. Robertson DSO OBE
Brig E. P. Townsend DSO OBE
Col N. F. B. Shaw DSO OBE
Col W. C. Walker DSO OBE
Second Bar to DSO
Maj Gen W. C. Walker DSO OBE
Bar to DSO
Lt Col W. C. Walker DSO OBE
DSO
Lt Col E. P. Townsend
Maj E. Gopsill MC
Lt Col D. L. Powell-Jones OBE
Brig A. G. Patterson OBE MC
OBE
Lt Col N. F. B. Shaw DSO
Lt Col J. A. R. Robertson DSO MBE
Lt Col W. C. Walker DSO
Lt Col E. P. Townsend DSO
Lt Col R. R. Griffith

Lt Col D. L. Powell-Jones MBE
Lt Col A. G. Patterson MBE MC
Lt Col W. B. Standbridge
Lt Col B. G. Hickey MC
Lt Col A. S. Harvey MC
Lt Col R. C. Neath
Lt Col C. J. Scott
Lt Col C. J. D. Bullock MC

MVO (5th Class)

Capt (QGO) Lalbahadur Thapa MC
Capt (QGO) Siriparsad Thapa
Capt (QGO) Partapsing Gurung
Capt (QGO) Tambasing Gurung
Capt (QGO) Manbahadur Gurung
Capt (QGO) Chandrabahadur Thapa
Capt (QGO) Pahalsing Thapa
Capt (QGO) Gambahadur Gurung
Capt (QGO) Dalbahadur Gurung
Capt (QGO) Manbahadur Tamang
Capt (QGO) Jaibahadur Gurung

MBE

Capt (KGO) Nandalal Thapa
Capt (KGO) Manu Gurung MC
Maj W. M. Amoore DSO
Maj N. E. V. Short MC
Maj A. G. Patterson MC
Maj (KGO) Lachhiman Gurung
Lt (KGO) Prembahadur Ghale
Maj (QGO) Nainasing Gurung
Maj W. D. McNaughtan
Hon Capt (GCO) Kharaksing Gurung
Maj (GCO) Kajiman Gurung
Capt (QGO) Jumparsad Gurung
Capt (QGO) Dhanbahadur Gurung MC
Capt (QM) W. J. Murray
Maj W. R. James
Maj R. N. P. Reynolds
Maj (QGO) Lokbahadur Thapa
Miss P. Castle WVS
Maj (QGO) Tejbahadur Gurung
Maj (QGO) Tambasing Gurung MVO
Maj E. D. Powell-Jones
Maj M. G. Hunt-Davis
Maj (QGO) Birkharaj Gurung
Maj Gopalbahadur Gurung
Mrs R. Carpenter WRVS

MC and Bar
Capt (QGO) Damarbahadur Gurung
Bar to MC
Maj A. J. D. Macdonald MC
Maj A. S. Harvey MC
Maj (GCO) Harkasing Rai MC IDSM MM
MC
Lt (KGO) Sherbahadur Gurung
Lt (KGO) Girmansing Thapa
Capt (QGO) Sarbajit Gurung
Capt (QGO) Lalbahadur Thapa
Capt R. E. W. Atkins
Lt (QGO) Dalbahadur Gurung IDSM
Capt J. A. Lys
Lt (QGO) Moti Gurung DCM
Lt (QGO) Ranbahadur Pun
Maj A. V. O. Robinson
DCM and Bar
21135027 Sgt Bhaktabahadur Thapa
DCM
21135018 CSgt Dalbahadur Gurung
21134008 CSgt Pahalsing Thapa
21135192 LCpl Pimbahadur Thapa
21145269 Sgt Tekbahadur Thapa
21134202 CSgt Ramsor Rai
MM and Bar
21136459 Sgt Amerbahadur Pun
MM
21134171 Sgt Chhetabahadur Pun
21135044 Sgt Jangbahadur Rana
21144597 Rfn Dhanbahadur Rana
21134243 Cpl Bombadahur Gurung
21135094 LCpl Rupsing Pun
21136040 Cpl Hembahadur Rana
21135183 LCpl Sobarnu Gurung
21143971 Sgt Bhupal Chhetri
21135154 Rfn Jitaram Ghale
21135100 Cpl Harke Thapa
21144049 Rfn Lasbahadur Gurung
21143884 Rfn Tekbahadur Roka
21144018 CSgt Dirgasing Thapa
21136088 Sgt Lale Gurung
21145599 Rfn Indrabahadur Gurung
21143373 LCpl Kabiraj Thapa
21133676 LCpl Kubirsing Gurung
21136277 Rfn Gagbahadur Pun

21134228 WO2 Lalbir Thapa
21136232 Rfn Balsing Gurung
21144387 Cpl Topbahadur Thapa
21136293 Rfn Narbahadur Thapa
21134129 Cpl Manbahadur Gurung
21135185 Cpl Indrabahadur Gurung
21136391 LCpl Indrabahadur Gurung
21145536 CSgt Karnabahadur Gurung
21138535 Cpl Balman Gurung
21144149 Sgt Pasbir Gurung
21138481 Cpl Bombahadur Gurung
21151962 Rfn Tulbahadur Thapa
BEM

21134320 Sgt Kishanbahadur Thapa
21143065 Sgt Bhokbahadur Gurung
21143005 CSgt Sherbahadur Rai
21134973 CSgt Uttamsing Gurung
21147965 CSgt Durgabahadur Sahi
21159031 LCpl Nandalal Gurung
      SSgt P. S. Brown Int Corps att 2/6 GR
22440565 SSgt R. J. Steppe REME att 1/6 GR
21153617 Sgt Narbahadur Thapa
21156191 Cpl Mohansing Gurung

FOREIGN DECORATIONS
BRUNEI
Most Blessed Order of Stia Negara
1st Class Maj-Gen W. C. Walker CB CBE DSO
2nd Class Brig A. G. Patterson DSO OBE MC
3rd Class Maj C. O. Fisher
      Maj (GCO) Tekbahadur Subba
      Maj J. S. Keen
4th Class WO2 Deobahadur Ale
Most Honourable Order of the Crown of Brunei
2nd Class Lt-Col A. S. Harvey MC
NEPAL
Gorkha Dakshina Bahu
4th Class Maj D. H. Houston MC
Tri Shakti Patta
4th Class Capt (QGO) Lilbahadur Gurung
PERAK
Distinguished Conduct Medal
21136190 LCpl Deobahadur Thapa

Colonels of the Regiment

1948–82

1926–1951	Field Marshal The Lord Birdwood of Anzac GCB GCSI GCMG GCVO CIE DSO LLD
1951–1961	Field Marshal The Lord Harding of Petherton GCB CBE DSO MC
1961–1969	Major General J. A. R. Robertson CB CBE DSO
1969–1974	Major General A. G. Patterson CB DSO OBE MC
1974–1978	Brigadier D. L. Powell-Jones DSO OBE
1978–1983	Brigadier Sir Noel Short MBE MC

Commanding Officers

1948–1982

Decorations given are those held during or awarded as a result of tenure in command.

1st Battalion

1947–48	Lt Col J. A. R. Robertson DSO OBE
1948–51	Lt Col E. P. Townsend DSO OBE
1951–54	Lt Col W. C. Walker DSO OBE
1954–56	Lt Col A. E. C. Bredin DSO MC
1956–58	Lt Col N. E. V. Short MBE MC
1958–61	Lt Col W. M. Amoore DSO MBE
1961–63	Lt Col W. B. Standbridge OBE
1963–66	Lt Col B. G. Hickey OBE MC
1966–68	Lt Col W. D. McNaughtan MBE
1968–69	Lt Col J. Whitehead MBE

2nd Battalion

1947–50	Lt Col N. F. B. Shaw DSO OBE
1950–53	Lt Col R. R. Griffith OBE
1953–56	Lt Col D. L. Powell-Jones DSO OBE
1958–59	Lt Col P. B. Winstanley MC
1959–61	Lt Col A. G. Patterson OBE MC
1961–63	Lt Col E. T. Horsford MBE MC
1963–66	Lt Col A. S. Harvey OBE MC
1966–68	Lt Col R. C. Neath
1968–69	Lt Col R. N. P. Reynolds MBE

6th Gurkha Rifles

1969–70	Lt Col R. N. P. Reynolds MBE
1970–73	Lt Col J. N. Kelly MC

1973–75   Lt Col C. J. Scott
1975–78   Lt Col M. J. F. Wardroper
1978–81   Lt Col C. J. D. Bullock OBE MC
1981–83   Lt Col P. D. Pettigrew

APPENDIX FIVE

Gurkha Majors

1948–1982

1st Bn

1948	Maj (KGO) Pahalman Gurung MBE OBI
1948–52	Maj (QGO) Lachhiman Gurung MBE
1952–57	Maj (QGO) Nainasing Gurung MBE
1957–62	Maj (QGO) Lalbahadur Thapa MVO MC
1962–65	Maj (QGO) Lokbahadur Thapa MBE
1965–69	Maj (QGO) Partapsing Gurung MVO
1969–69	Maj (QGO) Pahalsing Thapa MVO

2nd Bn

1948–50	Maj (KGO) Kulbahadur Gurung IDSM
1950–51	Maj (QGO) Nandalal Thapa MBE
1951–55	Maj (QGO) Hiralal Gurung MBE
1955–59	Maj (QGO) Dhanbahadur Gurung
1959–62	Maj (QGO) Jumparsad Gurung MBE
1962–66	Maj (QGO) Khusiman Gurung
1966–69	Maj (QGO) Amarbahadur Gurung

6th Gurkha Rifles

1969–71	Maj (QGO) Pahalsing Thapa MVO
1971–74	Maj (QGO) Kholal Ale
1974–78	Maj (QGO) Toyebahadur Chand
1978–82	Maj (QGO) Dalbahadur Gurung MVO
1982–	Maj (QGO) Jaibahadur Gurung MVO

APPENDIX SIX

Roll of Officers

1948–1982

Decorations given for those not on the Permanent Cadre are those awarded prior to and during service with the Regiment only.

Date of Joining	Name	Reason for becoming non-effective	Rank and Year
1931	Robertson J. A. R. CB CBE DSO	Retired	Maj Gen 1964
1932	Shaw N. F. B. CBE DSO	Retired	Col 1960
1937	Short N. E. V. MBE MC	Retired	Hon Brig 1964
1938	Amoore W. M. DSO MBE	Retired	Col 1966
1939	Lumley J. R.	Retired	Maj 1958
1939	Patterson A. G. CB DSO OBE MC	Retired	Maj Gen 1972
1940	James W. R. MBE	Retired	Maj 1958
1940	Brebner I. C.	Retired	Maj 1964
1941	Power P. F. MBE	Retired	Maj (QM) 1956
1942	Carroll C. S. F. MC	Transferred to 7GR	Maj 1964
1942	Morrison A. M.	Retired	Maj 1962
1944	McNaughtan W. D. MBE	Retired	Lt Col 1970
1944	Neath R. C. OBE	Retired	Lt Col 1979

1945	Walsh G. H.	Retired	Maj 1965
1948	Gahan H. A. B.	Transferred to Malay Regt.	Maj 1948
1948	Tregenza D. J.	Transferred to RAPC	Maj 1966
1948	Townsend E. P. CBE DSO	Retired	Brig 1961
1948	Pulley H. C. MC	Transferred to 7GR	Maj 1949
1948	Powell-Jones D. L. DSO OBE	Retired	Brig 1963
1948	Whitmell R. A.	Returned to R. Warwicks	Maj 1949
1948	Allen T. J. W. MC	Retired	Maj 1953
1948	Lorimer G.	Retired	Maj 1967
1948	Hutchison K. M.	Retired	Maj 1963
1948	Houston D. H. MC	Retired	Maj 1976
1948	Hickey B. G. OBE MC	Retired	Brig 1976
1948	Gopsill E. DSO MC	Transferred to 7GR	Maj 1949
1948	Organ D. C. MC	Retired	Lt Col 1976
1948	Allan L. W.	Returned to Essex Regt	Maj 1953
1948	Harvey A. S. OBE MC	Retired	Lt Col 1977
1948	Haynes D. V.	Left	Capt 1952
1948	Mann A. J. MC	Returned to KOSB	Capt 1950
1948	Fillingham J. A. I. OBE	Transferred to 7GR	Capt 1949
1948	Fisher A. J.	Transferred to 7GR	Maj 1966
1948	Taunton A. R. L.	Retired	Maj 1979
1948	Adams M. B.	Returned to Royal Scots	Capt 1950
1948	Axworthy W. N. A.	Returned to DCLI	Capt 1950
1948	Ross A. G. R.	Returned to Royal Leics	Capt 1950
1948	Douglas P. R.	Returned to Royal Northumberland Fusiliers	Capt 1949
1948	Studley F. J. MBE	Returned to Royal Berks	Capt (QM) 1954
1948	Henning G. W.	Left	Capt 1952
1948	Reynolds R. N. P. MBE	Still serving	Lt Col
1948	Pahalman Gurung SB MBE OBI	Retired	Hon Capt (GCO) 1951

1948	Kajiman Gurung MBE	Retired	Maj (GCO) 1955
1948	Barber R. L.	Killed motoring accident UK	Maj 1952
1948	Pockson M. H. DSO OBE	Transferred to 7GR	Lt Col 1955
1948	O'Bree C. P.	Retired	Maj 1970
1948	Pottinger L. E.	Retired	Maj 1970
1948	Barnes R. G.	KIA Malaya	Maj 1949
1948	Maxwell H.	Released	Lt 1948
1948	Moore D.	Transferred to 7GR	Capt 1949
1948	Proctor G. C.	Released	Capt 1949
1948	Cowan R. A. H.	Released	Capt 1948
1948	Mayman I.	Transferred to RASC	Lt 1950
1948	Vickers J. L.	Retired	Hon Maj 1963
1948	Golightly R. G. H.	Transferred to RA	Lt 1949
1949	Griffith R. R. OBE	Retired	Lt Col 1959
1949	Macdonald A. J. D. MC	KIA Malaya	Maj 1952
1949	Taggart A. B. MC	Transferred to 10GR	Lt Col 1959
1949	Morgan L. W.	Returned to Royal Sussex	Maj 1950
1949	Shaw P.	Returned to KSLI	Maj 1951
1949	Irving B. T.	Returned to Royal Norfolks	Capt 1952
1949	Bellers P. G. V.	Returned to Middlesex	Capt 1953
1949	Lys J. A. MC	Retired	Hon Lt Col 1976
1949	Robinson A. V. O. MC	Retired	Maj 1969
1949	Stone G. E. I.	Retired	Maj 1961
1950	Winstanley P. B. MC	Retired	Col 1971
1950	Hayward-Surry H.	Retired	Maj 1978
1950	Watson R. A. MBE	Returned to Yorks	Capt (QM) 1951
1950	Cardozo D. N.	Returned to S Lancs	Lt 1953
1950	Carter G. R. W.	Returned to RB	Lt 1951
1950	Scott C. J. OBE	Retired	Lt Col 1982
1950	Walsh D. J.	Retired	Maj 1976
1951	Walker W. C. KCB CBE DSO	Retired	Gen 1972
1951	Rees C. D.	Returned to RWF	Capt 1955
1951	Thomas A. T.	Returned to Queens	Maj 1953

1951	Atkins R. E. W. MC	Left	Capt 1958
1951	Hicks E. M. G.	Returned to Buffs	Capt 1955
1951	Haye J. H.	Returned to W Yorks	Lt 1953
1951	Gangabahadur Lama	Retired	Maj (GCO) 1964
1951	Bhimbahadur Thapa	Retired	Capt (GCO) 1954
1951	Turner H. G.	Returned to RUR	Lt (QM) 1952
1951	Eastap A.	Left	Lt (QM) 1955
1951	Anderson N. A. J.	Retired	Maj 1968
1952	Harkasing Rai MC IDSM MM	Retired	Maj (GCO) 1968
1952	Fisher C. O.	Retired	Maj 1970
1952	Brighton E. G.	Left	Capt 1959
1952	Davie J. H. B.	Left	Capt 1955
1952	Gray A. K.	Left	Capt 1963
1952	Paterson P. F.	Left	Lt 1955
1952	Ferguson J.	Returned to Manchesters	Lt (QM) 1954
1952	Whorlow T. G.	Retired	Maj 1969
1952	Bellers T. G.	Transferred to 2GR	Maj 1962
1952	Knights J. R.	Retired	Maj 1976
1952	Croce J. N.	Left	Lt 1953
1953	Hart G. N. B.	KIA Malaya	Capt 1955
1953	Ashby H. W.	Left	Capt 1957
1953	Clee J. B. B.	Retired	Lt Col 1981
1953	Prismall R.	Left	Capt 1965
1953	Tweed M. A.	Left	Lt 1958
1954	Bredin A. E. C. DSO MC	Returned to Dorsets, (Staff)	Brig 1956
1954	Wilkie E. H.	Retired	Maj 1970
1954	Barber C. J.	Returned to Royal Berks	Capt (QM) 1956
1954	Hedges W. J.	Transferred to 7GR	Lt 1956
1954	Adshead D. R.	Retired	Maj 1971
1954	Peters I. S.	Left	Capt 1962
1955	Fitzgerald J. A. D.	Returned to Royal Norfolks	Capt 1958
1955	Robinson T.	Died in Oxford	Capt (QM) 1960
1955	Hariprasad Dewan	Retired	Maj (GCO) 1970
1955	Allsop G. B.	Left	Lt 1958
1955	Dunlop C. V. C.	Left	2Lt 1956
1955	Willson J. J.	Retired	Maj 1970
1955	Hill R. B.	Left	Lt 1958
1955	Smith P. J.	Left	2Lt 1957

1955	Moore W. H. MBE	Died	Capt (DoM) 1957
1956	Bromet J. A. R.	Left	Lt 1958
1956	Goater E. L.	Left	Capt 1965
1956	Wakeham A. C.	Left	Capt 1962
1956	Ward A. C.	Killed motor-cycle crash	Lt 1958
1956	Wilson R. W. F.	Retired	Maj 1976
1956	Hamilton P. J.	Left	2Lt 1957
1956	Churchill C. ff	Left	2Lt 1957
1957	Wooley H. MBE	Retired	Maj (QM) 1965
1957	Tekbahadur Subba	Retired	Maj (GCO) 1971
1957	Robeson P. B. H.	Retired	Maj 1970
1957	Hill E. A.	Retired	Maj 1969
1957	Osmond A. P. K.	Left	2Lt 1958
1957	Taylor M. I.	Left	2Lt 1958
1957	Garside R. R.	Left	2Lt 1959
1958	Stanley J. A.	Left	2Lt 1959
1958	Hill D. B.	Left	2Lt 1959
1958	Beacham R. E.	Left	2Lt 1959
1958	Rawlings R. L.	Left	2Lt 1960
1959	Whitehead M. L.	Still serving	Lt Col
1959	Wallace F. H.	KIA Borneo	Lt 1963
1959	Carron R. P.	Left	2Lt 1960
1959	Walker J. R. H.	Left	2Lt 1961
1959	Simons D. K. S.	Retired	Capt 1970
1959	Coldicott R. J.	Retired	Hon Maj 1971
1959	Lowe R. H.	Retired	Hon Maj 1971
1959	Macleod Q. N. B.	Retired	Maj 1970
1960	Pettigrew P. D.	Still serving	Lt Col
1960	Stamper W. H.	Left	2Lt 1961
1960	Burge H. L.	Retired	Maj (DoM) 1970
1960	Murray W. J. MBE	Retired	Maj (QM) 1973
1960	Furtado J.	Left	Lt 1962
1960	Powell-Jones E. D. MBE	Transferred to 7GR	Lt Col 1981
1960	Gordon I. S.	Left	Capt 1969
1960	Channing M. D.	Left	Capt 1967
1960	Riley F.	To General List	Capt (QM) 1962
1961	Standbridge W. B. OBE	Retired	Col 1970
1961	Horsford E. T. MBE MC	Retired	Lt Col 1968

1961	Akalsing Thapa	Still serving	Maj
1962	Craft J. W. R.	Returned to DLI	Capt (QM) 1965
1962	Keen J. S.	Retired	Maj 1969
1962	Walker N. H.	Left	Capt 1969
1962	Hunt-Davis M. G. MBE	Transferred to 7GR	Maj 1976
1962	Conlin J. W.	Left	Lt 1965
1962	Boucher S. M.	Returned to Buffs	Capt 1965
1963	Whitehead T. E. K.	Still serving	Maj
1963	Herring J. C. G.	Returned to Kings	Capt 1966
1964	Mackinlay J. C. G.	Still serving	Maj
1964	Gopalbahadur Gurung MBE	Still serving	Maj
1965	Shoesmith R. W.	Left	Capt 1976
1965	Neville-Davies K. G.	Left	Lt 1967
1965	Gray K. B. W.	Left	Lt 1967
1965	Powell-Jones C. P.	Left	Capt 1974
1965	Streather H. R. A. MBE	Returned to Glosters	Maj 1966
1965	Prevett A. F.	Returned to 14/20H	Capt (QM) 1967
1966	McIntyre D. W. M.	Left	Lt 1967
1966	Briggs D. H. McK	Still serving	Maj
1966	Huggard R. C. H.	Left	Lt 1970
1966	Mackenzie I. W.	Left	Lt 1970
1966	O'Bree B. M.	Still serving	Maj
1966	Anderson J. A.	Still serving	Maj
1966	Duncan R. H.	Still serving	Maj
1966	Hewitt F. S.	Returned to Inniskillings	Capt 1967
1966	Larthe de Langladure C. H. A.	Left	Lt 1971
1967	Hornel G. C.	Left	Lt 1970
1967	Doodson H. N.	Transferred to R HAMPS	Lt 1971
1967	Pettigrew D. J.	Left	Lt 1971
1967	McGuire W. E.	Retired	Maj (QM) 1976
1967	Pett R. A.	Returned to QLR	Capt 1969
1967	Godfrey D. A.	Retired	Maj 1976
1968	Whitehead J. MBE	Retired	Brig 1980

Year	Name	Status	Rank
1968	Gouldsbury P. P. A.	Still serving	Maj
1968	Morris T. R.	Left	Lt 1971
1968	Freeland C. J. S.	Left	Lt 1971
1968	Lewis D. W.	Left	Capt 1981
1969	Corrigan J. G. H.	Retired	Maj 1980
1969	White D. J. H.	Returned to Kings	Capt 1971
1969	Davies G. L.	Still serving	Maj
1969	Scrase M. W.	Returned to QLR	Capt 1971
1970	Aitken-Quack P. N.	Left	Lt 1972
1970	Kelly J. N. MC	Still serving	Lt Col
1971	Oakden H. N.	Retired	Lt Col 1980
1971	Bhagwansing Thapa	Still serving	Maj
1971	Neale A. J.	Left	Lt 1974
1971	Bushell I. P.	Still serving	Maj
1971	Anderson C. F.	Left	Lt 1975
1971	Merrick A. W.	Returned to R Sigs	Lt 1973
1972	Fonfe F. D. C.	Left	Capt 1978
1972	Titley J. D.	Still serving	Maj
1973	Pearce R. J. A.	Left	Capt 1981
1973	Harman M. A.	Still serving	Capt
1973	Lewis P. N.	Left	Capt 1979
1974	Underhill P. A. T.	Still serving	Maj
1974	Manikumar Rai	Still serving	Capt
1974	Richardson-Aitken R. F.	Still serving	Maj
1974	Wathen N. C. J.	Left	Lt 1977
1974	Fletcher P.	Left	2Lt 1975
1974	Patterson H. W. G.	Returned to RGJ	Capt 1977
1975	Wardroper M. J. F.	Still serving	Lt Col
1975	Creamer R. A.	Retired	Maj (QM) 1979
1976	Gordon-Creed N. A. D. F.	Still serving	Capt
1977	Lys G. D.	Still serving	Capt
1977	Jarvis D. A.	Still serving	Lt
1978	Cox R. C. B.	Transferred to 2GR	Maj (QM) 1979
1978	Bullock C. J. D. OBE MC	Still serving	Lt Col
1979	Mulvaney D. M.	Still serving	Capt (QM)
1979	Baugh D. M.	Still serving	Lt
1979	Griffith A. P. M.	Still serving	Capt
1979	Peters E. A.	Still serving	Lt

1979	Cameron-Hayes J. C.	Returned to 14/20H	Capt 1980
1979	Degraaf P. J.	Returned to RAR	Capt 1981
1980	Edmunds H. C.	Left	Lt 1982
1980	Marwick-Smith J. R.	Still serving	Lt
1980	Bradly S. C.	Left	Lt 1982
1980	McClean R. W. K.	Returned to KOSB	Capt 1981
1980	Sutton N. J.	Returned to Gordons	Capt 1982
1980	Shirreff A. R. D.	Returned to 14/20H	Capt 1981
1981	Buckeridge J. C. M.	Still serving	2Lt
1981	Bulbeck D. E. P.	Still serving	2Lt
1981	Birch J.	Left	2Lt 1981
1982	Athill J. A.	From 3 RGJ. Still serving	Capt
1982	Latter C. M.	Still serving	2Lt

APPENDIX SEVEN

Attached Officers

1948–1982

RAMC

1st Bn

1952–53	Capt J. S. V. Surman
1953–55	Lt D. M. Cunningham
1955–56	Capt J. D. Jefferies
1956–57	Capt I. W. Young
1958–59	Capt I. Esslemont
1959–61	Capt A. Slora
1961–62	Capt B. Duff
1962–63	Capt D. N. Wood
1964–66	Maj G. Somerville
1966–70	Maj R. J. O. Smith

6th Gurkha Rifles

1970–71	Capt N. K. I. McIver
1971–74	Lt Col G. Somerville
1974–76	Capt P. L. Thomas
1976–79	Capt B. J. Heap
1979–81	Capt J. Luby
1981–82	Capt J. W. A. Moore
1982–	Capt A. J. Reidy

2nd Bn

1950	Capt W. McIllwraith
1950–51	Capt D. Cook
1951	Capt J. Macleod
1951–52	Capt J. McCormick
1952–54	Capt B. D. Fairgrieve
1954	Capt G. P. T. Barclay
1954–55	Lt M. R. Nowell
1955–57	Capt D. O. Jones
1957–58	Capt A. R. Worthington
1958–59	Capt M. R. Bicknell
1959	Lt A. A. Hooper
1959–61	Capt G. F. Berry
1961–62	Capt T. Astin
1962–64	Capt I. W. W. Anderson
1964	Capt P. A. North
1964	Capt D. Townsend
1964–67	Capt K. J. H. Southern
1967–68	Capt J. O'Donnell
1968–70	Capt C. R. Bradshaw

RAPC

1st Bn

1957–60	Maj F. J. P. Wardell
1960–62	Capt B. M. Bowen
1962–65	Capt A. G. Perrin
1965–69	Maj A. R. Tawney

2nd Bn

1959–69	Maj E. L. Burk

6th Gurkha Rifles

1969–70	Maj E. L. Burk
1970–73	Capt E. Stroud
1973–78	Capt F. A. Gates
1978–79	Capt A. B. Holmes BEM
1979–82	Maj R. J. Corbett
1982–	Capt J. W. G. Smith

RAEC

1st Bn

1958–62	Lt G. D. Stroud
1962–67	Capt D. W. Matheson
1967–69	Capt C. C. Baker

2nd Bn

1962–65	Capt J. M. Owen
1965–67	Capt P. J. Ball
1967–69	Capt J. M. Roberts

6th Gurkha Rifles

1969–70	Capt J. M. Roberts
1970–73	Capt N. B. Thomas
1973–76	Capt D. J. Wickens
1976–78	Capt K. S. Boulter
1978–81	Capt G. Bradbury
1981–83	Capt A. A. Rawlings

APPENDIX EIGHT

WRVS

1948–1982

1st Bn

1949–50	Miss E. M. How
1950–51	Miss P. Makepiece
1951–53	Miss M. O. Potton
1953–54	Miss Palmer
1954–55	Miss J. Hill
1955–56	Miss N. Rushforth
1956–57	Mrs L. Gilbert
1957–67	Miss P. Castle MBE
1967–69	Mrs R. Carpenter

2nd Bn

1949–52	Miss D. Nunn
1952–53	Miss L. Ackery
1953–59	Miss E. M. Davidson
1954–57	Miss D. Price
1957–64	Miss E. McAvoy
1964	Miss J. Howell
1964–66	Mrs A. Walters
1966–68	Miss K. Batey
1968–69	Miss J. Thompson

6th Gurkha Rifles

1969–83   Mrs R. Carpenter MBE
(Not with Bn in Church Crookham 1977–79)

Index

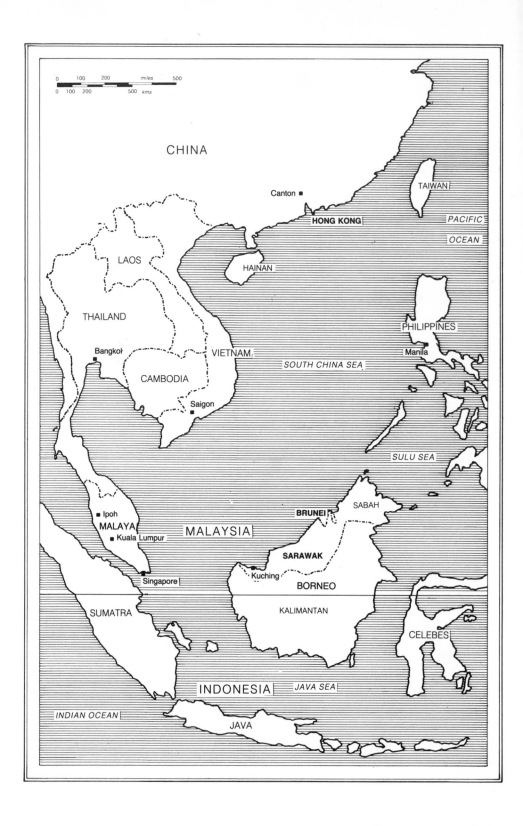